The Southern Way

The regular volume for the Southern devotee

Kevin Robertson

Issue 55

www.crecy.co.uk

© 2021 Crécy Publishing Ltd
and the various contributors

ISBN 9781800350267

First published in 2021 by Noodle Books
an imprint of Crécy Publishing Ltd

New contact details
All editorial submissions to:
The Southern Way (Kevin Robertson)
'Silmaril'
Upper Lambourn
Hungerford
Berkshire RG17 8QR
Tel: 01488 674143
editorial@thesouthernway.co.uk

All rights reserved. No part of this book may be reproduced or transmitted in any form or by any means electronic or mechanical, including photocopying, recording or by any information storage without permission from the Publisher in writing. All enquiries should be directed to the Publisher.

A CIP record for this book is available from the British Library

Publisher's note: Every effort has been made to identify and correctly attribute photographic credits. Any error that may have occurred is entirely unintentional.

Printed in the UK by Short Run Press

Noodle Books is an imprint of
Crécy Publishing Limited
1a Ringway Trading Estate
Shadowmoss Road
Manchester M22 5LH

www.crecy.co.uk

Front cover:
A well-stained BR Standard 2-6-2T No. 82024 on a train of 18 hoppers at Feltham, in September 1965. The single disc above the right hand buffer indicates Feltham as the destination but not where from.
Graham Smith courtesy Richard Sissons

Rear cover:
Having featured part of the London area within the covers it was only appropriate we continued the theme here. Balham station in a mix of green and later BR colours. The foot crossing between the platforms was once commonplace but apart from the very occasional location where there is no bridge or subway, such things are nowadays strictly 'verboten'.
Douglas Twiball/Transport Treasury

Title page:
Past times...once upon a time views such as here were commonplace and we took it all for granted.

Contents

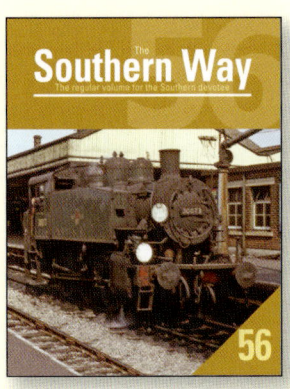

Issue No 56 of THE SOUTHERN WAY
ISBN 9781800350274
available in October 2021 at £14.95

To receive your copy the moment it is released, order in advance from your usual supplier, or it can be sent post-free (UK) direct from the publisher:

Crécy Publishing Ltd (Noodle Books)

1a Ringway Trading Estate, Shadowmoss Road, Manchester M22 5LH

Tel 0161 499 0024

www.crecy.co.uk

enquiries@crecy.co.uk

Introduction 4

'Ten Times T' 6
   The London, Chatham & Dover
   Railway 0-6-0 Tanks
   Mike King

The Railways of Dulwich..12
   A Personal Perspective
   Michael Rowe

The S. C. Townroe Archive – in Colour 34
   Part 4

Down to Earth Part 4..45
   Ex-SECR Stock
   Mike King

Superheater Development on the
   Southern Railway 57
   John Harvey

Buriton Chalk Pits..63
   John Perkin

Belmont Hospital Bridge 69
   George Hobbs

Branch Line Society Tour 72
   7 March 1959

Down to Earth 81
   A Postscript
   John Burgess

Christmas Mails and Parcels Traffic 83
   A Retrospective Review Part 1
   Richard Simmons

Introduction

One of the things that is a joy in compiling *Southern Way* (or indeed any other book or an article) is how I can almost always guarantee to learn something new. Perhaps I should qualify that slightly by saying, perhaps, I re-learn, having previously forgotten the information over past years. Here too is something some might call sad, for on one of my recent afternoon walks I found myself reciting engine names, class types, stations and even the names of signal boxes along particular routes. Some might say I am slightly touched ('sad' is the popular term also used) but I would say the opposite as it helps keep the grey matter functioning and it is also amazing how little snippets of 'useless' information that have emanated from that railway interest have come in useful in general conversation and, of course, quizzes over the years. That does not mean I talk trains all the time! Indeed I suspect, like many, my knowledge of UK geography owes much to that interest in railways from long past.

I think too the opportunity to 'visit' places I was never fortunate enough to see in their heyday through the medium of photography has helped. It took some years but I came to realise there was far more to the network than the limited geographical area I was permitted or, should I say, had the opportunity to explore as a boy.

On not dissimilar lines, I was recently asked by another publisher if I was prepared to speak to their marketing department to explain what that railway interest of ours is all about. (Makes me wonder if the right person was in the right job perhaps, but that was not for me to question.) So how does one explain a railway interest to an individual whose likely perception is either an old man in a grubby anorak or someone called 'Kevin' – I blame my parents for that – taking numbers at a station?

Well this Kevin never did take numbers. He remembers a few but I never did have a notebook and pencil, I just wanted to. Instead I just sat and watched and sometimes managed to cajole a short footplate trip or signal box visit but mostly it was just watching.

And to answer the question as to what we do; well some of us like a particular railway or era, others like to work to recapture the past at a museum or on a heritage line, others make models or run them, usually a combination of more than one of these. Others buy books and journals to reacquaint themselves with their interest and for the sake of good old nostalgia. And I say good luck to you all.

Ours is a harmless persuasion shared by a wide cross section, from Lords of the Realm to professionals, military, naval and air force men, engineers, white collar, blue collar, pop stars, in fact a pretty fair cross section of humanity. It is rarely, and should not be, the 'be all' of life, but instead part of it. It is a harmless pastime whilst there are also some incredibly clever individuals involved with amazing retentive memories. So really it is not much different to a person having an interest in say stamps, motorcycles, aircraft, sea shells, Greek mythology – or whatever. The principal difference to me is that whilst some interests are socially acceptable, 'trains' is not so much.

All these answers were politely listened to but I still don't think I managed to get through. In the end what really worked was when I turned the tables and asked, "What does the firm you work for actually do?" The answer was obvious, "They produce (train) books". I responded by suggesting the employer might also be interested in the subject as well, which of course produced an affirmative response. And here was the nub, railways may be an interest to some, but it can also be employment, a business, or just a hobby to others.

All too often it is easy to criticise a person or what they do or stand for, for the simple reason we do not really understand. I just hope I managed to turn a few things around in the mind of one person at least.

I conclude this missive in breaking one of my traditions in singling out a particular article within this issue – well, the start of one really. This is Richard Simmon's gargantuan piece on Christmas Mail and Parcels Traffic. Yes it might be long, yes it is of necessity going to have to be spread over at least two issues and yes it does have considerable detail which perhaps all might not feel they want to read – but they should. If you want to have some idea of what running the railway was like before the days of mass communication and information at the touch of a button then this is an example of how it was done. It is also the sort of information that might so easily have been lost. As it is similarly said by our supposed leaders in parliament "I commend *this article* to the house".

Kevin Robertson

Opposite top: **Another change of circumstances. Around a century before this view was taken Nine Elms terminus was in its heyday; here though it is in goods department use, but to be fair, even on the Southern, goods was a worthwhile and remunerative traffic for many years.**

Bottom: **Still on the topic of infrastructure – and in this respect we are almost continuing the subject (although not the locations) from the previous issue. Here we have Haven Street in BR days but with the cast signage almost looking as if it would have been at least two decades earlier.**

WAY OUT
6
NO

HAVEN STREET
WAY OUT
WARNING
STOP LOOK & LISTEN
BEFORE CROSSING THE LINE
SOUTHERN RAILWAY
BEWARE OF
TRAINS

'Ten Times T'
The London, Chatham & Dover Railway 0-6-0 Tanks
Mike King

Well, almost ten – as, although we have ten pictures, we do not have pictures of all ten examples! Indeed, they may be considered a rather forgotten class as the last was withdrawn in 1951 but, like all things William Kirtley, they performed their duties with quiet efficiency. The first two, LCDR Nos. 141 and 142, emerged from Longhedge Works in December 1879 and were notable in that they were the first tank locomotives built wholly by the company. Both were employed at Battersea yard – mainly on shunting duties. The remaining eight examples followed between 1889 and 1893 – LCDR Nos. 143-150 – but they did not appear in numerical order. There were some detail differences between these and the original pair – and some of the details of Nos. 141/2 were later brought into line with the others. Most were allocated to Battersea shed (Stewarts Lane in post-1933 SR days) and used on local goods and shunting turns, and later for empty stock workings into and out of Victoria. Regular passenger work did not commence until 1894 when they took turns on inner city passenger trains, for which purpose they were equipped with Westinghouse brakes and LCDR smokebox and bunker destination board brackets.

No. A600 at Longhedge on 1 June 1929, in SR lined black livery. Originally LCDR No. 141, it ran from December 1879 until December 1936. Purchased for £350 by R. Frazer & Company, it departed northwards on 29 January 1937 and was later sold to Richard Evans & Co of St. Helens, Lancs., being used at Haydock Colliery until September 1958 – easily outlasting all other members of the class with a total lifespan of 79 years. The headcode disc indicates Stewarts Lane to Victoria; almost certainly on empty coaching stock. In the right background may be glimpsed an ex-SECR gunpowder van – a not very common class of wagon. The picture is possibly by Henry Casserley although (unusually for him) it is not marked as such on the reverse, as the writer has another different photograph of the engine taken on the same day – an SLS shed visit maybe?

'Ten Times T'

The other 1879 original, A601, at Ashford on 22 July 1933, soon after withdrawal. Both this and No. 600 had slightly shorter frames, so the cabs were a little more restricted for space and coal capacity was fractionally less than the others. The boilers were also of a different pattern, but both were reboilered in July 1908 making them look practically identical to the later examples. *H. C. Casserley*

No. A602 inside Ashford Works on 27 July 1931, with some repainting having already commenced, but despite the date, the A prefix and original Southern number remain. The date 2/31 is chalked on the smokebox (date of boiler overhaul, perhaps?), while presumably the number 680 refers to the boiler itself. Note destination board brackets still on the smokebox door – no doubt unused for some time previously. *H. C. Casserley*

After the initiation of joint operating – the companies were never actually amalgamated until they became part of the Southern in 1923 – with the South Eastern, they had 459 added to their Chatham numbers (the SER's highest numbered loco was 458), becoming SECR Nos. 600-609 over the period 1900-3 but continued to work from Battersea shed. Come 1914, the requirements of wartime saw them start to move around the SECR system, with Nos. 600/7 going to Dover and 601 to Folkestone. At least six travelled to Boulogne during 1915 while the other four later went to Rouen – all receiving ROD dark grey livery but were renumbered 5600-9 to avoid confusion with other ROD locomotives carrying the same numbers. They must have been well-received as the last was not repatriated until 1919. On return to the SECR they were distributed as follows: Bricklayers Arms Nos. 602/5/6, Folkestone No. 608, Dover Nos. 600/3 and the other four returned to Battersea. However, apart from No. 607, which did a spell at Maidstone, the rest had migrated back to Battersea

A later view of LCDR No. 143, now as SR 1602 in unlined black livery at a murky Stewarts Lane on 15 April 1937. Lamp in the usual position to indicate Stewarts Lane-Victoria. This was the final locomotive in BR stock, withdrawn in wartime black with Bulleid lettering at Reading on 7 July 1951 – being cut up on 1 September following. Mileage is recorded as 988,764 but just how this was computed for a shunting locomotive is hard to imagine. Perhaps some sort of daily formula was applied. (From Alastair Wilson – "I always understood that there was a formula, but that it depended on the size of the Yard, so it varied from place to place.")
W. Leslie Good

A view across the South Western main line to the west of Wimbledon station, showing the Engineer's yard, probably in late 1929. No. A603 has arrived with a ballast train from the Wimbledon-St Helier-Sutton line, which was then under construction and opened in January 1930. Loco A600 was also employed on this work. An ex-LBSCR 20-ton brake van to SR Diagram 1576 – SR No. 55917, is coupled next to the locomotive while all other wagons visible are of ex-Midland Railway origin. The signals just visible over the rooftops apply to the joint LSW/LBSC line towards Merton Park, thence onward to Tooting and Mitcham junctions. *The Lens of Sutton Association*

'Ten Times T'

by December 1920.

Renumbering as Southern Railway Nos. A600-9 began in March 1924 and was completed in July 1926 – the first overhaul; A605 receiving passenger green livery – but all the others were painted lined goods black. A605 conformed with the others from its next general repair, while post-1932 overhauls saw the numbers increased by 1000 (ie. 1600-9) and lining dispensed with, although not all locomotives would have seen these changes before withdrawal. During early Southern days all returned to Battersea shed and were much used on empty carriage workings into and out of Victoria. However, in 1929/30 Nos. A600/3 were used on the construction of the new

Loco A604, also at Ashford on 22 July 1933. Like No. A601 in a previous picture, it was withdrawn from service in May 1933 – the rods are already off – and offered for sale but as no buyer emerged, second thoughts must have prevailed for instead it was overhauled, to re-appear from Ashford Works in July 1934, now numbered as 1604. Withdrawn again in January 1939, like the proverbial cat it had nine lives and was repaired and returned to duty in the following November – to serve out the war and after at Herne Hill sidings. Loco A608 may just be seen to the right. *H. C. Casserley*

In the guise of SR No. 1604, former LCD No. 145 continued to serve at Herne Hill until 1949 and was then rusticated to Reading shed in January 1950 – ever a dumping ground for old SECR locomotives. There it was employed on local shunting duties and ballast trains as far as Guildford. Minor repairs at Ashford in April 1950 saw it return to Reading in BR livery as No. 31604 – the only one to carry BR numbering. It is seen at Reading shed in June 1950, still carrying its LCDR smokebox door destination brackets. Withdrawal came soon after – on 4 November. It was cut up at Ashford on 13 January 1951. The bunker of No. 1602 is just visible to the right and this managed to live on at Reading for another 12 months. *W. Gilburt, courtesy Ian Wilkins*

Wimbledon-Sutton line – opened in January 1930.

Withdrawals commenced in 1932 with A605 and by the end of 1936 all except Nos. 1602, 1604 and A607 had been taken out of traffic. Several were then offered for sale although by no means all found buyers. Nos. 1600/8 went to the rolling-stock dealer R. Frazer of Hebburn-on-Tyne and both ended their days serving collieries in the north of England. No. A607 was set to be purchased by Messrs. Wake of Darlington, but the deal fell through and instead the loco was taken into Ashford Works in August 1938, to reappear as departmental locomotive No. 500s and destined for Meldon Quarry, near Okehampton. There it replaced another South Eastern loco – Manning Wardle 0-4-0 saddle tank No. 313. This gave the West Country its only sighting of a former London, Chatham & Dover Railway locomotive, where it remained until December 1948. Its replacement was a more commonplace loco for the area –

No. 500s fulfilled the Meldon Quarry duty with occasional visits to either Okehampton or Exmouth Junction sheds – and perhaps back to Eastleigh on occasions – until December 1948, when the firebox failed to pass an examination. Sent to Eastleigh for repair, a period of storage followed until the loco was condemned on 1 October 1949. It is seen soon after, in a line of stored locomotives – on 29 October. It was last noted at Guildford on 18 December, en-route to Horley for breaking up which took place in a siding alongside the Brighton main line on 7 January 1950. Several other locomotives were also cut up at this location during 1949/50. *A. E. West*

A nice broadside view of A607 – probably outside Victoria and soon after repainting in Southern lined black livery, as the SECR plate remains on the bunker. The repainting date was June 1925. Withdrawn in December 1936, it was stored for a time at Stewarts Lane while a sale was negotiated, but this came to nothing and it journeyed to Ashford Works before re-emerging as departmental loco 500s in September 1938 – being dispatched to Meldon Quarry as the resident shunter. *Author's Collection*

G6 tank DS3152.

Of the other two SR survivors, No. 1604 was taken out of traffic in January 1939 but was not broken up – instead it was stored following Government instructions in the event of hostilities – and was repaired and reinstated at Ashford Works in November 1939. It was then sent to shunt Herne Hill sidings, where it joined No. 1602 and both locomotives continued in gainful employment until mid-1949. Both were then stored for a short period at Stewarts Lane before being sent to Reading shed for local shunting duties there. In April 1950 No. 1604 visited Ashford Works for minor repairs and emerged numbered 31604 – the only one to ever carry a BR number – but despite this was withdrawn soon after. No. 1602 was retained at Reading until July 1951, but may not have seen much use during its final year.

No. A608, again at Ashford on 22 July 1933 – along with A604 seen behind. This loco also returned to service for another spell, being condemned on 19 December 1936. This was the other loco to be sold to R. Frazer, it was resold on to the Wallsend & Hebburn Coal Company in July 1937 – serving out its time at the optimistically named Rising Sun Colliery until August 1948. After three more years of dereliction, it was broken up by Bowrans of Newcastle. So, despite official SR records indicating to the contrary, half of the class managed to see in the 1950s - just. *H. C. Casserley*

The Railways of Dulwich
A Personal Perspective
Michael Rowe

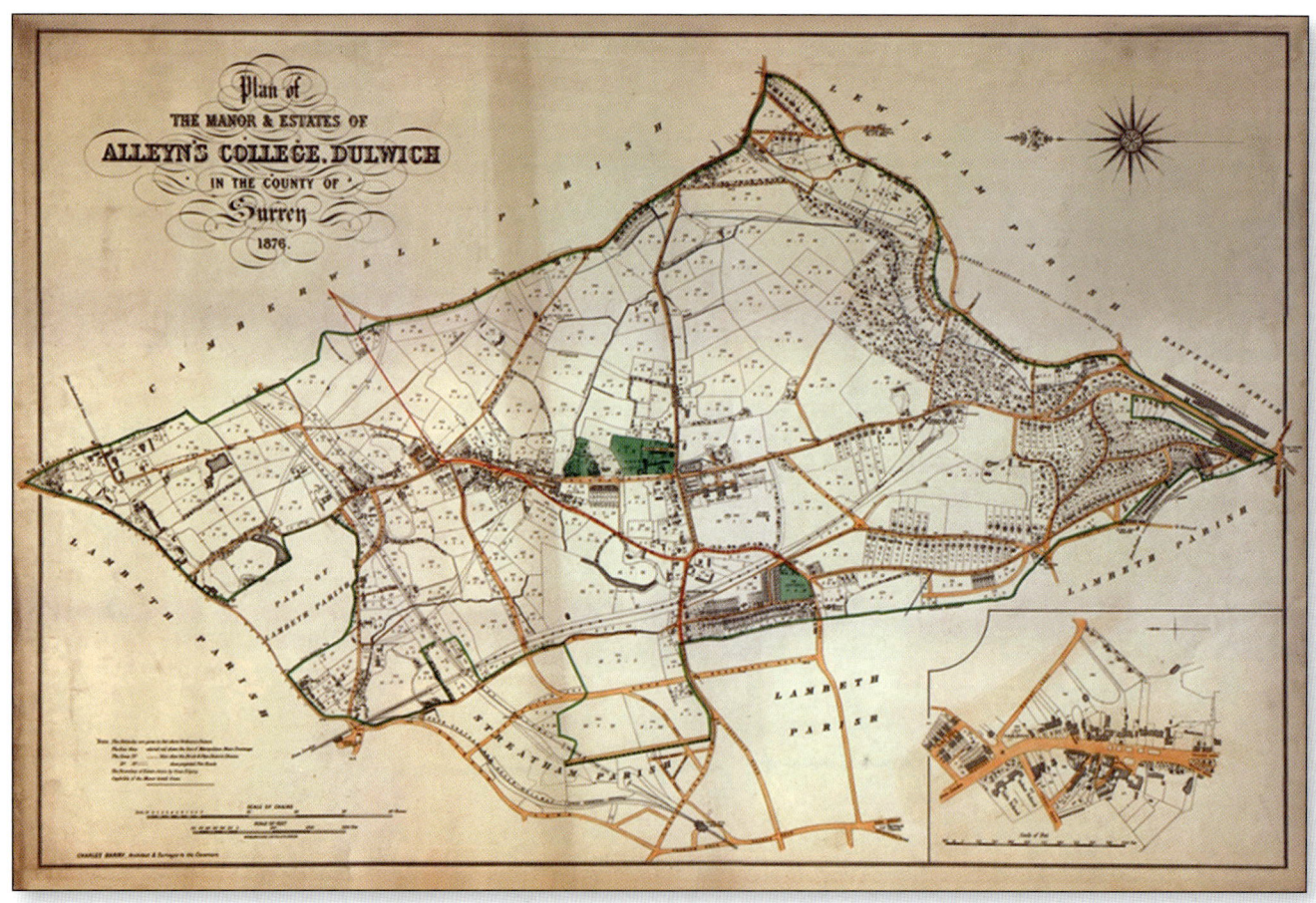

Map of the Dulwich Estate circa 1876. This shows the four routes that opened during the preceding decade.

You know how it is: you plan a journey, it has a purpose; a clear objective, or so you think. You buy your ticket and board the train and everything seems set fair. But then as you progress your focus shifts as diversions occur and familiar certainties shift and drift. In this particular case I was initially enthused by the prospect of writing about a place that featured significantly through the early phases of my life and which, on the increasingly rare occasions I now return, still engenders a sense of belonging and charged pleasure. Walking through its green spaces, or sipping a pint of ale in one of its hostelries, my consciousness inevitably filters back to the period when I was growing up, or living there: seeking out memories of those with whom I communed. So geography and personal history were without doubt drivers but there was something else. I had always assumed the coming of the railway to this locality would have been viewed as a trial, a scar on the landscape, something to be resisted. But once I began my research I was surprised to discover my presumptions were, to a considerable degree, misconceived: so, I kept digging and then, finally, there was a wartime incident that occurred no more than two hundred yards from my former bedroom and of which I first learned almost thirty years ago. Details were scant: if I was going to write about the history of the railways in Dulwich, then finding out what happened that winter evening in Knights Hill Tunnel would have to be a key element.

During the mid-1960s my parents, with me in tow, moved from a flat on a dowdy Victorian estate in Walworth to a spacious terraced house in West Dulwich. The Rowe family therefore arrived in Dulwich precisely a century after railways first ran across the Dulwich Estate. By the age of fourteen my interest in the railway was already deeply entrenched, although at that time it still largely focused on the decline and imminent extinction of steam locomotives on British Railways metals.

A Charles Barry Junior original: North Dulwich Station. *All images by the Author unless otherwise stated*

Later I broadened my appreciation of what I loosely define as railway archaeology, concentrating in particular on the impact railways have on landscape and communities, the residue of which can remain long after services cease and tracks are lifted.

Dulwich is a village within Greater London, which like Hampstead, Blackheath and Wimbledon has retained some of its early character despite the rapid spread of urbanisation during the late nineteenth and early twentieth centuries. However, in the case of Dulwich there are particular circumstances which enabled it to resist this urgent expansion and what proved to be, for most metropolitan localities, the inevitable consequences: expanding industrial activity and housing for the associated workforce. When my parents bought their house they acquired a leasehold interest and their superior landlord was the Governors of Alleyn's College of God's Gift: otherwise, the Dulwich Estate. Within ten years of their acquisition, with the coming into force of the Leasehold Reform Act 1967 and with me training to be a solicitor, I found myself acting for not only my parents, but many of their neighbours in connection with the enfranchisement of their homes, whereby they acquired their freeholds subject to the terms and conditions of a scheme of management, approved by the High Court in 1974. That Scheme vested the management and control of the common areas of the Dulwich Estate with its Governors, who to this day are entitled to levy an annual maintenance charge upon the householders. The Governors can still impose restrictions on alterations to houses of which they once owned the freehold.

Those negotiations brought me into regular contact with the Estate Office in Dulwich Village and the College's solicitors, Druce and Attlee, a firm that had been retained by Dulwich College for over two hundred years. I acquired an understanding of Dulwich's history and how the village and its environs had been managed in the recent past and I made assumptions, as I was bound to do, of how the emergence of the railway had encroached upon Dulwich and of the resistance its progress must surely have met. I was aware of how many landowners responded during the Railway Mania of the mid-nineteenth century. It is obvious to anyone who studies a map of the Dulwich Estate that the four railway lines that did eventually traverse the Estate only did so along its periphery and well away from its heart.

Dulwich Village is approximately four miles south of Charing Cross. Geographically it sits in a shallow basin akin to a wide soup bowl, with higher ground on all sides. To the north there is Herne Hill, which extends northward to become Denmark Hill, close to where John Ruskin resided at the very time the railway was extending its tentacles out from the metropolis. To the east the land rises towards One Tree Hill and Honor Oak; to the south there is the ridge of Sydenham Hill, which sits on top of a steep escarpment and along which The Crystal Palace stood from 1854 until its destruction by fire in 1936. Today its dominant feature is the BBC television transmitter which rises towards the sky, a poor imitation of Eiffel's Tower. To the west is Knights Hill, a lump of land that separates Dulwich from the ancient settlements of Lower (now West) Norwood and Streatham. In their midst sits Dulwich, a flat and still relatively green enclosure, which until the early twentieth century remained primarily a rural district, with as its focal point its seat of learning: Dulwich College.

Dulwich is not mentioned in the Doomsday Book. Allen Galer in his history, 'Norwood and Dulwich', published in 1890, comments that "Dulwich has always been a quiet, little, picturesque place." He suggests its name may have derived from the medieval spelling 'Dilwysshe' which refers to a Mr De la Wyk, who is said to have owned considerable lands in Camberwell during the reign of Henry I.

In his history of Dulwich College, William Young makes no mention of De la Wyk, but suggests Dulwich means: "Village in the valley." Both Galer and Young, whose book was published a year earlier in 1889, agree on the village's chain of ownership, which over eight centuries has but four links. In 1127 the Crown vested the Manor of Dulwich in the Priory of Bermondsey, where it remained until the Reformation. In 1539, due to the administrations of Thomas Cromwell, the land reverted to the Crown and in 1544 Henry VIII sold the Manor to Thomas Calton at a price of

£609. 18s. 2d. Thomas's son Sir Francis Calton succeeded his father as owner of the Manor, but it seems he was feckless and by the early years of the seventeenth century had accumulated considerable debts. Much of the Manor of Dulwich had been mortgaged and Francis was in the market, seeking a buyer. Enter Edward Alleyn, an actor, theatre manager and contemporary of Shakespeare.

Edward was not one of William Shakespeare's close associates, indeed he became a rival. Having once worked with the Bard, Alleyn set up 'The Fortune', a direct imitation of Shakespeare's Globe Theatre, and clearly he made a success of that venture. In 1605 he first acquired part of Sir Francis Calton's landholdings, allowing the latter to pay off his debt to Sir Robert Lee, and then exchanged articles of agreement to buy the remainder of the Manor, which he purchased for a consideration of £4,900. During the following years further parcels of land were added and by the time the school was established in 1619, for which a Royal Charter had been granted, the extent of the current Dulwich Estate was intact and has largely remained unaltered. The Estate, subject to the enfranchisement of individual houses and the terms of its scheme of management, remains in the ownership of the Governors of Dulwich College, who are the trustees of the successor to the original charitable foundation created by Edward Alleyn, for his 'College of God's Gift'.

In his essay, 'Planning a London Suburb: Dulwich 1882 – 1920' (published in 1994), Bernard Nurse records that in 1892 the Dulwich Estate comprised an estimated 1,167 acres. It extended from Denmark Hill in the north to Crystal Palace in the south and from Forest Hill in the east to Knights Hill in the west. He goes on to state that in 1901 Dulwich "had the lowest population density of any registration sub-district within 6 miles of the centre of London." He also notes: "Dulwich was exceptional in having one estate… of considerable size and situated in a block, making it possibly the largest compact estate in the County of London. The landowner was therefore potentially in a stronger position to determine development while subject to less influence from neighbours." These were factors that were clearly in play during negotiations with individual railway companies.

So, what of the railways? Before considering historical issues relating to Dulwich College and my family, I should address the core subject. As I noted earlier the railway did not arrive in Dulwich until the 1860s. This was comparatively late in relation to the early establishment and subsequent rapid development of railway routes and businesses from the centre of London outwards: that had begun in the mid-1830s. One reason for this delay was that Dulwich, safe in its green valley, was not a destination in itself. Nor, historically, was it a staging post on the way to anywhere else. Allen Galer observes, "The nearest high road lay two or three miles off, passing through Streatham and Croydon and the road that traversed Dulwich simply led to the still smaller village of Sydenham."

In his study, 'Victorian Suburb (a study of the growth of Camberwell)', H. J. Dyos observes that the development of routes through the Dulwich area prior to 1860 was minimal and slow because the major railway companies were primarily interested in developing main lines to distant destinations such as Dover, Brighton and Southampton. The first established routes to these destinations passed either to the west of Dulwich in the case of the south coast towns, or for the Kent routes, to the north. Neither the South Eastern or London, Brighton & South Coast railway companies appeared keen to act as "suburban pioneers" by seeking to establish routes where there was no proven or obvious demand. In the 1850s Dulwich remained largely rural: there was the village with its chapel, art gallery and the school, plus some fine Georgian houses strung along the thoroughfare known as Dulwich Village, but there had not yet been any significant development.

Collegiate and corporate crests on the road bridge opposite North Dulwich Station. This bridge carries Red Post Hill over the South London Line. Similar crests appear on all of the bridges carrying that line through Dulwich.

The Railways of Dulwich

Circumstances began to alter with the re-location of The Crystal Palace following the Great Exhibition of 1851. The glass and steel structure was re-erected along the crest of Sydenham Hill from where it would dominate the sky line of South London until its fiery destruction eighty years later. The Palace was clearly going to become an attraction and the railway companies responded accordingly. First out of the blocks was the West End and Crystal Palace Railway, which quickly established a new route from Victoria. However, that route passed through Streatham and Lower Norwood, before reaching what was to become Crystal Palace Low Level station via Gypsy Hill. It did not therefore benefit the inhabitants of, or provide access to, Dulwich.

However, the London, Chatham & Dover Railway had by 1860 obtained approval for the construction of three metropolitan lines, all of which would run to, or through, Herne Hill. Herne Hill Station opened in August 1862 with services to Victoria via Brixton. It now sought to establish an alternative route to Dover via Bromley and became the first company to secure land over the Dulwich Estate. This acquisition took place in 1861 and the line from Herne Hill through Dulwich Station (later West Dulwich), Sydenham Hill and then onwards through a lengthy tunnel to Penge East, opened in 1863. The line runs along the western and southern fringes of the Estate.

In the meantime the LBSCR was keen to establish a route from Peckham Rye to Streatham, which would link up with established lines to Wimbledon and on to Sutton. Eventually a route via East Dulwich, North Dulwich and Knights Hill Tunnel to Tulse Hill was authorised and this hugs the northern and western boundaries of the Estate. The city terminus for this line, over which services began in 1868, was London Bridge, which was also the point of departure for the third route that crossed the north-eastern extremity of the Dulwich Estate: this was the Crystal Palace and South London Junction Railway, which opened in 1865 and ran from a connection with the LBSCR at Cow Lane Junction near Peckham Rye, via Nunhead and Lordship Lane, finally reaching the northern escarpment of Sydenham Hill and a rather grand terminus, which would become Crystal Palace High Level.

Although these three lines ran along the periphery of the Dulwich Estate, with the exception of the route to Crystal Palace, they did not do so unobtrusively. The Dover main line runs largely along an embankment almost entirely comprising earth works, but including a section of three hundred yards that sits upon a low ornamental brick and stone viaduct. This feature was constructed at the direction of the Governors of Dulwich College, "at considerable additional expense".

The original route proposed by the LBSCR, and for which parliamentary approval was obtained, sought to avoid any passage across the Estate by skirting around its north-western border. The LBSCR were apparently mindful of the demands Dulwich College had imposed upon the LCDR. As will be discussed later the original route was abandoned for a more direct line south to Tulse Hill and the College was able to ensure the infrastructure for the new line was both substantial and ornamental.

A southbound service has departed North Dulwich and is heading for Tulse Hill with its leading coaches crossing Village Way.

The route to Crystal Palace, in so far as it passed over the Dulwich Estate, did so largely through cuttings and via two tunnels. The passage of down trains from Lordship Lane ran through Dulwich Wood alongside Cox's Walk and then through Crescent Wood Tunnel, beyond which Upper Sydenham Station was subsequently opened, before reaching Crystal Palace via Paxton Tunnel. This route, perched upon the western slope of Sydenham Hill, was and indeed still is, in so far as traces remain, picturesque.

So, the Governors and their consultants, Charles Barry Junior, who had succeeded his esteemed father Sir Charles Barry as surveyor and architect to Dulwich College in 1858, and Charles Druce, the College's solicitor, found themselves locked in negotiations with several railway companies as the 1860s approached. They had understandable concerns to conserve the distinct rural characteristics of the Estate, but I now appreciate this was a moment they had doubtless been anticipating with relish and expectation for several years. It was not the case that they viewed the coming of the railway as a threat to Dulwich; it was instead an opportunity for the Governors to secure the finance they required to advance their charitable activities into a new era.

As previously stated, Edward Alleyn created his charitable foundation in 1619. The original objects were the education of the young and the accommodation of the elderly. The school was to provide education for twelve poor scholars and also for fee paying pupils. The welfare of the elderly was to include the provision of almshouses for "poor brothers and sisters" chosen from four parishes with whom Alleyn had been closely connected: St Botolph's Bishopsgate, St Saviour's Southwark, St Giles Cripplegate and the local parish of Camberwell, within whose boundaries his manor was located. Alleyn intended his charitable foundation should "endure and remain forever". Indeed, there are to this day almshouses close to the centre of Dulwich Village and they stand adjacent to the seventeenth century chapel.

In an attempt to ensure the longevity of those objectives, Alleyn enshrined his foundation by way of a series of detailed statutes that were granted a Royal Charter and which appear to have been very prescriptive. They also contained a number of quirks and eccentricities. For instance, one statute required the Master of the College, who was in effect the principal of the charity, to have the same surname as the original benefactor. So, for each generation a man with the surname of Alleyn, or Allen, had to be identified and appointed to succeed his predecessor.

By the nineteenth century the activities of foundations such as Alleyn's were subject to supervision by the Crown's Charity Commissioners. In the 1840s the manner in which Dulwich College was being administered and the standard of education its scholars were receiving was a cause of concern to several parties. An association was established by former pupils, who complained about low levels of academic achievement and the brutality of certain masters. They stated insufficient pupils were obtaining access to university, or to apprenticeships. The parishes were also disgruntled regarding the general governance of the charity and their concerns were reinforced by criticisms expressed by the Archbishop of Canterbury, who had been petitioned by the Old Scholars' Association. He called for significant alterations to be made to the management of Dulwich College.

The process by which those alterations came about is detailed by William Young in his extensive history of Dulwich College, which was published in 1889 and deals with the period from inception until 1858, when a new constitution was initiated. The Charity Commissioners began an investigation into the governance of Alleyn's Foundation in May 1854. The Master and his advisers knew what was coming and it does seem they had begun to prepare their response to the criticisms they faced. In June that year the Master, solicitor and various other parties were "examined" by the Commissioners, as evidence was sought in respect of the College's revenues and governance. Other witnesses included, "several of the poor brethren and sisters" who were occupying the almshouses and alumina of the school.

One former scholarship boy, Robert Farmer of Mount Street, Lambeth, stated in his deposition that he believed "the lands of the College were underlet". Charles Druce in his evidence remarked that when it came to lettings the advice of the College's surveyor was always followed. He went on to say he believed the Estate was not underlet and referred to two recent leases of land in Dulwich Wood. In his text Young confirms that during the early 1850s there were negotiations between Dulwich College and The Crystal Palace Company and also with two individuals (who I believe were developers), relating to "the Sydenham portion of the Estate". Young says these negotiations were handled by Sir Charles Barry and Charles Druce and resulted in "a very large increase to the College Revenues". For reasons I shall explain, I suspect these disposals evidence a change of attitude by the Governors and their recognition that change was not only coming, but needed to be embraced.

Following the gathering of depositions, Charles Druce advised the Charity Commissioners that, "In the present state of the enquiry [the College is] not prepared to say whether [it] will send any scheme". In other words, the College was leaving it to the Charity Commissioners to propose the scope of the alterations to be made to Edward Alleyn's foundation. The Commissioners duly issued proposed heads of a new scheme to the College in May 1855 and by the end of that year, the Master of the College had signed off his response, which contained objections and suggestions. Young also reports that: "By this time the College had begun to consult the Commissioners in respect of matters concerning its management, for at this meeting a letter from the Secretary was read leaving it to the discretion of the College to assent or dissent from the East Kent Railway; and it was resolved that the College should dissent, notices thereof being accordingly signed by the Master." The scope of the proposal being advanced by the railway company is not disclosed.

During 1856 and 1857 the Charity Commissioners presented two Bills for its new scheme to the House of Lords. The first was lost when Parliament was dissolved in March 1857, but after detailed debate and consideration at committee stage, during which the College and other interested parties were represented by counsel, the Bill finally passed and became law

Threading its way through the village in the valley: the railway crosses Village Way.

on 25 August 1857. The scheme approved by Parliament was largely that which the Commissioners had originally proposed, although amendments had been added or removed as the Bill progressed.

The new scheme provided for the dissolution of the existing "Corporation" and the establishment of an upper and a lower school, "with foundation scholarships attached to the latter and exhibitions to the former." The Upper School became Dulwich College and the Lower School was to become Alleyn's School, which moved to its current site in Townley Road during the 1880s. The net revenues were to be divided into four parts of which three were to be allotted to the purposes of the schools and one to the maintenance of the almspeople. Governance of the scheme was to be vested in fifteen governors, several of whom would be elected by the beneficiary parishes, with the remainder to be nominated by the Court of Chancery, one of whom had to be a resident of Dulwich. The Master of the Upper School and the headmaster of the lower school would be appointed by and answer to the governors.

Charles Druce, when asked if he approved of the scheme, replied, "My opinion is that it is a very large departure from the Will of the Founder." However, he also commented, when asked if the scheme was beneficial, "Undoubtedly I think it is a very great improvement on the present management." A major alteration was that under Alleyn's original foundation the charity's income was required to be divided equally between the educational and the charitable branches. The new emphasis on education was an invitation for the Governors, in their new guise, to seek finance for the replacement of the dilapidated buildings from which the existing school was then operating and to expand the scope of the educational institution beyond the confines imposed by Alleyn's statutes.

So it was that on 31 December 1857 Alleyn's charitable foundation was dissolved after almost two hundred and forty years and the new Scheme was established.

My assertion is that the new Governors, their role now strengthened with powers defined in less antiquated terms, were keen to embrace the coming of the railway and were not, as I had previously thought, resistant to its intrusion: this is supported by what happened to Knights Hill. This prominence sits to the west of Dulwich Village, rising from Croxted Road, formerly Croxted Lane, alongside which, from 1862, the LCDR main line, elevated upon its embankment, ran parallel to that highway. Knights Hill was not, at the precise moment the new charitable trusts took effect, part of the Dulwich Estate. Maps of the estate which appear on the Dulwich Society website clearly show that in 1850 the western boundary was immediately to the east of Croxted Lane. Knights Hill is not marked on the 1850 map, but the land to the west of Croxted Lane is shown to lie within "Streatham Parish".

Knights Hill (which should not be confused with another hill of the same name and no more than a mile to the south-east, with a thoroughfare which runs from near West Norwood Station up to Crown Point in Upper Norwood, from where it continues east as Beulah Hill) is a mound of clay. Both hills are said to have acquired their name from Henry Knight and his family, who in the sixteenth century held, as tenants, large parcels of land in Norwood and Streatham. It is said Henry Knight operated a tile kiln on Knights Hill from which he supplied forty-six thousand roof tiles in 1537 for use on houses built on old London Bridge. As I can from personal experience attest, Knights Hill soil is heavy clay: the house my parents bought was in Rosendale Road, approximately half way between the top and the foot of its eastern flank. Our rear garden ran uphill and beyond its fence there stretched a broad patchwork of allotments, which in the 1960s reached up towards a row of trees strung along the summit.

During the eighteenth century Knights Hill came into the ownership of Edward Thurlow, later the First Baron Thurlow, a distinguished lawyer who served as Lord Chancellor for fourteen years in the governments of four different prime ministers. Thurlow built a house – Thurlow Park – in what was then Lower Norwood and acquired tracts of land in Norwood and Streatham to supplement his original acquisition. In 1785 he bought the Knights Hill Estate from the Duke of St Albans. Edward Thurlow died in 1806 and forty years later his heirs sold Knights Hill to another lawyer, Charles Ranken. It is reasonable to assume this sale would have been advertised and indeed, there may well have been an auction. Whatever the circumstances, Dulwich College does not appear to have displayed any interest in acquiring the land.

In William Young's history of Dulwich College there is the following footnote:

> "With money received by the old Corporation for the land taken by the Crystal Palace, the Knights Hill property, 59 acres 2 roods and 20 perches, was purchased in 1859 for £13,000 from the devisees of Mr Charles Ranken. This was part of the estate of Lord Thurlow (died 1806) and was bought from his heirs by Mr Ranken in 1846. From exchange, sales to railways, etc, the area of this property is now 52 acres 2 roods and 27 perches."

So, my submission is that Dulwich College bought Knights Hill purely for the purposes of achieving the best outcome from their pending negotiations with the railway companies. Presumably Mr Ranken's devisees did not avail themselves of the necessary intelligence regarding the likely routing of a railway line south through Dulwich to Tulse Hill, or they would surely have forced a harder bargain. A consequence of the College purchasing Knights Hill was that the land on which my parents' house was later to stand was brought within the Dulwich Estate, and in one of the appendices to Young's book he notes, "it should be placed on record that the present Turney Road, and the portions of Rosendale Road and Thurlow Park Road which are on the manor, were made at the expense of the railways when they first passed the property". It is pleasing to reflect that the segment of Rosendale Road where my parents lived for fifty years was first constructed at the behest of the LBSCR.

A section of the viaduct that runs between Village Way and Croxted Road, looking south from beside the bridge over Burbage Road.

The Railways of Dulwich

The viaduct that carries the South London Line viewed from Turney Road. Dulwich Cricket Club provides the foreground. The spire of St Paul's Church on Herne Hill is visible.

As the 1860s commenced, the passage of private bills through Parliament for the establishment of railway lines across parts of the Dulwich Estate began in earnest. Those private acts would include rights of compulsory purchase, but the landowners were entitled to be compensated for their loss and for a large estate owner such as Dulwich College there was considerable scope to impose restrictions and stipulations binding upon the railway companies. It seems the LBSCR was justified in its caution about proceeding with a route that crossed the Estate.

A bill was lodged in Parliament and passed during 1862 for a route that avoided the Estate. However, there was subsequently a change of heart and a further Act was passed the following year – The South London Act – which provided for a line to Tulse Hill that took a more direct route south west across part of the Dulwich Estate. This was the preferred option for the LBSCR, provided a satisfactory connection could be made with the LCDR at, or near, Tulse Hill; assuming, of course, suitable agreement could be reached with the Governors of Dulwich College. The connection with the LCDR was achieved: it is the line that runs south from Herne Hill and to the west of Knights Hill before reaching Tulse Hill Station. This connection was approved by yet another act of parliament that passed in 1863 and at that time its route passed along the very edge of the Estate.

By then the Governors had concluded an agreement with the LCDR regarding the line from Herne Hill to Penge East. They no doubt felt emboldened by the satisfactory conclusion of those negotiations, because John Howard Turner in his publication, 'The London, Brighton & South Coast Railway', records that reaching agreement with the Governors regarding the South London Line took far longer than the company originally contemplated. Agreement was finally signed off in May 1864.

In anticipation of achieving that agreement, the LBSCR obtained yet another Act for the new line, which passed on 29 July 1864. In a schedule to this legislation there are set out the terms and conditions agreed between the railway company and the Governors in consideration of the College giving up ten acres of land. It would appear several of those acres were part of the land purchased from Charles Ranken in 1859. Some of the more significant restrictions and stipulations were:

- All bridges to be constructed on land formerly within the Dulwich Estate would be completed to the design of the Governors' architect and those carrying the railway would require "wing walls or screens extending to 150 feet on each side of the bridge, opening to correspond with the design of the parapet".

- A suitable station for passenger and parcel traffic "and all other traffic usually carried by passenger trains only" was to be erected on a site immediately adjoining Red Post Hill, but the station was not to be used for other forms of traffic without the consent of the Governors. In addition all trains each way, except for "Express Trains", would be required to stop at the new station.

- The Governors' architect was commissioned to design the station and all bridges, viaducts, and other buildings and works "connected with the Railway on the Estate". The architect was to be paid by the Company a commission of five per cent of the cost of the works. Responsibility for the construction works, in as far as they formed part of the "Railway Requirements", would lie with the "Engineer of the Company", although all works on the Estate were to be carried out to the reasonable satisfaction of the "Surveyor of the Estate".

- All boys attending the Governors' schools would be carried by the Company's trains "at one half of the annual fares payable by persons or children of corresponding ages travelling by trains or carriages of the same class".

The route of the South London Line following departure from Peckham Rye enters the Dulwich Estate south of East Dulwich Station (which is outside the Estate), progresses into a cutting to reach North Dulwich Station where substantial brick retaining walls were constructed, before emerging from beneath the road bridge under Red Post Hill. Shortly the line crosses Village Way before bestriding a redbrick shallow-arched viaduct that carries the line past Herne Hill Velodrome, over Burbage Road and then beside a series of sports grounds. The viaduct ends as the line passes over the Dover main line, following which it continues along a widened embankment that once contained the advance sidings of Knights Hill Goods Depot, before passing over another road bridge at Rosendale Road and entering Knights Hill Tunnel. This short tunnel has a handsome brick portal at its southern end, which is clearly visible from Tulse Hill Station. The northern entrance is not visible from Rosendale Road or neighbouring land and that is probably why it appears to lack the adornments Charles Barry prescribed for the southern facade. Tulse Hill Station lies several hundred yards to the south of the tunnel.

For such a short section there is a great deal of content when it comes to infrastructure. Furthermore, the College clearly had every intention of blending the railway furniture with the Estate in order to minimise loss of amenity. To achieve this, the

This July 2020 view shows the replacement span of the bridge carrying the South London Line over Rosendale Road. In the left foreground is the retaining wall of the 1891 bridge that formed part of Knights Hill Goods Depot. In the distance the high brick arch of the bridge that carries the former LCDR line linking Herne Hill and Tulse Hill can be seen.

The Railways of Dulwich

Governors had just the man to precisely design the style of structures they desired: Charles Barry Junior. He had succeeded his father Sir Charles Barry in 1858. Sir Charles was a renowned architect who had as part of a distinguished career overseen the rebuilding of the Palace of Westminster following its destruction by fire. He was noted for his major contribution to the use of 'Italianate' designs, especially the use of the 'Palazzo' as the basis for country houses, city mansions and public buildings. Sir Charles had been appointed architect to the Dulwich Estate in 1830: he died two years after his retirement.

Charles Barry Junior, like three of his brothers, followed his father into the profession and was clearly influenced by his father's Italian leanings. This is immediately apparent from the buildings and structures he designed in Dulwich. He was responsible for all the railway stations: North Dulwich, West Dulwich, Lordship Lane and Sydenham Hill. He also designed Crystal Palace High Level. Lordship Lane, which was demolished in the 1950s, following closure of the Crystal Palace High Level Branch, was by all accounts an impressive building which was elaborately ornamented, with two steeply gabled roofs above walls of red and black stone. The main building at West Dulwich has the appearance of an Italian villa, but the jewel is surely North Dulwich Station.

The fascia overlooking Red Post Hill has three arches with Tuscan Doric columns and sculpted keystones. The timbered landing and stairways down to the two platforms, with their high brick walls, are still intact. The building is Grade II listed and largely unaltered. As John Howard Turner succinctly states, "the station at North Dulwich is magnificent".

It was not however just the stations that impressed. As the railway crosses the Estate there are a number of road bridges. Barry designed them with a flair and adornment that is rarely matched across the whole British railway estate. As a small boy, long before we moved to Rosendale Road, we would come to Dulwich regularly, because my father played cricket and football at The Borough Polytechnic sports ground in Turney Road. We usually travelled by bus, changing at Herne Hill, although sometimes we caught the train. Either way, we then took a number 3 bus up to the Turney Road stop on Croxted Road. That short bus journey passed beneath two bridges: first a bridge over Croxted Road, close to the junction with Norwood Road, which stood on land that once formed part of the Dulwich Estate. This bridge carries the LCDR link line from Herne Hill to Tulse Hill. Like the other bridges,

Contrast the July 2020 image of Rosendale Road with this 1953 view, which shows all three of the railway bridges that once crossed Rosendale Road, with the defunct 1891 structure in the foreground. The two LBSCR bridges still retain their original ornate ironwork. A class C2X locomotive is shunting in Knights Hill Goods Depot. The absence of motor vehicles is startling and the trees rising behind the LCDR bridge stand in Brockwell Park.
R. C. Riley/The Transport Treasury

The Lairdale housing estate now occupies the site of Knights Hill Goods Depot. All that remains of the Depot is a brick retaining wall, the parapet of which rises above a terrace of lock-up garages and which once stood beside the vehicular entrance and roadway that rose from Rosendale Road up to the Depot. The parapet is clearly visible and at its far end is one of the stone abutments of the 1891 bridge, upon which the weathered remains of corporate crests are still visible. The facing wall of the northern side of the bridge can be seen on the far side of Rosendale Road from where the sidings once extended northwards through an area that is now heavily wooded.

it boasted ornamental ironwork and a badge denoting the letters 'AC' for Alleyn's College. The facades were independent of the beams that carry the railway. There were also four chamfered columns that were non-structural and purely decorative. The second bridge a short distance further up Croxted Road carries the South London Line: there are four brick buttresses either side on which three badges appear. One is the badge of the LBSCR, another bears the letters 'AC' with the date the bridge was built – 1866 – and finally there is the badge of the Dulwich Estate.

After we got off the bus, we would turn into Turney Road, where the entrance to the sports ground was a hundred yards away. Between us and the gate was a bridge carrying the Dover main line and this too had a plaque in the centre of its balustrade, which confirmed the date of construction, 1863. This bridge had broad, graceful, green iron pillars and all of these bridges were a cut above the utilitarian structures that carried the railway over the cluttered streets around the Elephant & Castle, which I walked by or under every day.

The highest bridge over-sails Burbage Road at the start of the section that sits astride the viaduct. Metal badges bearing the designs of Alleyn's College and the LBSCR are sited on each high brick pier. Other notable bridges are those carrying the same line over Village Way and under Red Post Hill. The latter has four identical cast iron painted panels fixed to the brick parapets and once again the badges of the LBSCR, Alleyn's College and the Dulwich Estate are displayed. The southern portal to Knights Hill Tunnel, which faces towards Tulse Hill Station, also benefits from the grandeur and ornamentation of its design, which comprises patterned brickwork and a crest bearing the arms of Dulwich College. The northern portal is largely obscured save for a fleeting glimpse as trains enter the tunnel via a brief cutting. Many of the bridges have been the subject of replacement and strengthening works in recent years. As a consequence some of the original features have been lost, but the overarching ambition of Barry's designs can still be discerned.

As the dust settled following the construction of the new railway lines, the Governors had good reason to reflect with satisfaction on the outcome they had achieved. In total, one hundred acres were transferred to the railway companies for a return of £1,000.00 per acre. They also procured transport facilities that would benefit the Estate for many decades ahead. From the cash received, the Governors proceeded to build a new school, designed of course by Charles Barry Junior, a splendid red brick, Italian-styled concoction of spires, towers and palazzos in College Road, which became the famous public school and which can be glimpsed from trains passing between West Dulwich and Sydenham Hill. The new College, which was opened by the Prince of Wales in 1870, is said to have cost £80,000.00 and its alumni include: P. G. Wodehouse, Raymond Chandler, Denis Wheatley, Ernest Shackleton and Nigel Farage.

So within twenty-five years Dulwich College was transformed from a failing institution with poor governance and decrepit premises to a modern, forward looking charity, with a sumptuous palace for its school with surplus capital. The railways played a significant role in this transfiguration, although elements of the conservative ethos of Alleyn's original foundation would survive.

As I have hinted, it was no accident my parents chose to buy a house in Dulwich. That they did so was an immense achievement and the consequence of twenty years of diligent saving. Jim and Joan married in 1946 and had lived in their Pullens Estate flat in Penton Place since just before I was born in 1951. My father had first visited the sports ground in Turney Road in 1936 when he joined Borough Polytechnic Technical School. He would go to Dulwich to play sports and from numerous conversations down the years, it was evident that spending time each week in those semi-rural environs contrasted greatly with the smut and murk of the Elephant & Castle. When he left school he joined the Polytechnic's sports club and so his weekends were largely spent in Dulwich. He was not a man who waxed lyrical, but several times down the years he told me how much he appreciated the view of the spire of Dulwich College Chapel, which peeps above the tree-line on the far side of the first eleven cricket pitch. Moving to Dulwich and being near the sports club, with my mother acquiring the garden she had long craved, were key objectives for my parents. Its realisation must have seemed almost utopian.

Despite the advent of the railways, housing development across the Dulwich Estate did not significantly accelerate during the final decades of the nineteenth century. The land sold in the 1850s near Dulwich Wood had been developed to create what Bernard Nurse describes as, "an exclusive suburb of large detached houses with extensive grounds... erected on the slopes of Sydenham Hill for the wealthy families who sent their sons to the College." In the early years of the twentieth century the Governors were anxious to attract further families to Dulwich who would support their schools and in particular wealthy families whose sons might attend the public school. The Governors successfully fought off various proposals for the introduction of electric tramways offering cheap fares, which in their view would have made the area more attractive to working class families. It was said the value of property "along such routes would be lowered".

In 1901 Camberwell Metropolitan Council, clearly incensed and frustrated by the Governor's stance, threatened to exercise compulsory purchase powers, so that housing for working people could be built on the Dulwich Estate. The Governors petitioned the Charity Commissioners and successfully argued such action would be detrimental to the interests of their, by then, three schools and proffered their own development scheme. This was approved and led to the construction of estate cottages in Dekker Road, close to the centre of the village.

An overview of the South London Line and Knights Hill Goods Depot taken from the northern slope of Knights Hill in May 1953. The parapet wall seen in the previous image is clearly visible on the right beyond the furthest queue of wagons, as are the two road bridges. The signal box stands alongside the up line at the far end of the bridge carrying the through lines. A 4-SUB unit is approaching on the down line from North Dulwich, while a locomotive appears to be shunting in the sidings on the north side of Rosendale Road. The entrance to the tunnel is behind the photographer's left shoulder. The roofs and chimneys of the Peabody Trust Estate are clearly visible immediately to the left of the through lines, with Herne Hill rising beyond.
R. C. Riley/The Transport Treasury

Gradually new houses were constructed along the main roads of the Estate and by 1913 a mixture of large Edwardian style terraced and semi-detached houses had been constructed in Rosendale Road. It was one of those houses my parents bought in 1965. The irony is they were precisely the kind of people considered undesirable by the Governors sixty years earlier; working class and with no prospect of sending their teenage son to Dulwich College. A further irony was they bought the house from a family named Wilson and Mr Wilson was a master at the College.

My recollections of the railways around Dulwich are many and varied. The former LCDR main line ran along one of the boundaries of the sports club. In the 1950s there was a steady flow of boat trains, coastal expresses and local services to and from Orpington, passing by. During that first decade of my life, long distance trains, including the *Golden Arrow*, were steam hauled. Several of my parents' sporty friends had children of a similar age and we, in the words of one of those childhood companions, "ran free". In a corner of the sports ground, immediately beneath the railway embankment, stood a tall tree with a small clearing to its rear. This became the setting for many imaginary sieges or skirmishes, or if girls were involved, more domestic scenarios. As we played the trains provided a constant backcloth with the bark of a Bulleid exhaust drowning our voices, or wisps of steam momentarily enveloping us. By the time I entered my second decade, all services on this line were electric and by then I was becoming more interested in sporting activities. But, you were always aware of trains running along the grassy, tree lined embankment and, when viewed from the heart of the sports ground, of the mound of Knights Hill that rose beyond the railway.

Once we moved in, my horizons regarding Dulwich rapidly expanded. There was much to explore. I quickly discovered that just two hundred yards from our house, where the South London Line crossed Rosendale Road, there was a goods and coal depot. This had required a second road bridge to be built shortly before it opened in 1892. The Governors had imposed restrictions forbidding the installation of goods sidings at North Dulwich Station, so the depot was built on railway land that was by then outside the Estate. Development of land for housing around Herne Hill and West Norwood had progressed rapidly and indeed immediately to the west of Knights Hill and the railway in Rosendale Road was the site of the Peabody Trust's first suburban development. The land was acquired in 1889 and by 1902 four blocks of flats had been constructed. During the next four years a further eighty-two cottages were added.

Knights Hill providing the backcloth and a final view of the bridges over Rosendale Road. In May 1953, Class C2X No. 30525 is easing out of the sidings and appears poised to cross over onto the main line towards North Dulwich. However, the signal on the up through line is 'off', so this may instead be a shunting manoeuvre. *R. C. Riley/The Transport Treasury*

Peabody's housing stock was intended for "working people in secure employment". Couples who applied for tenancies had to produce evidence they were married and a superintendent acted as the landlord's agent to ensure compliance with tenancy terms and conditions.

The goods yard was owned by a "foreign" company, the London & North Western Railway. It was one of several such yards operating on what is now Southern Region territory. Indeed, in Walworth we had lived a few hundred yards from the Midland Railway's coal depot off Walworth Road. By the time we arrived Knights Hill Depot was in its final, lingering death throes. It was located on the down side of the line from North Dulwich and had two connections to the main line. One was on the north side of Rosendale Road from where the depot tracks then crossed that road over the new bridge. The other access point was a trailing connection that joined the down main line shortly before the entrance to Knights Hill Tunnel. The yard was flat, with a high brick retaining wall that was set against the hillside and comprised five sidings in the main depot, with two further sidings along the northern access spur on the opposite side of Rosendale Road.

This was predominantly a coal depot. In 1906 seven businesses were registered as occupiers, with five designated as coal merchants, one a builders' merchant and the other having no trade specified. The yard was served at one time by a daily working from Peckham Rye Depot (a joint LNWR and Midland facility), with the return roster proceeding on from Peckham to Lillie Bridge Sidings on the West London Joint Railway. After the Second World War the demand for coal began to decline and by the early 1960s there were just two goods trains a week. Vehicular access to the Depot was on to Rosendale Road through a gate immediately adjacent to the 1891 bridge.

By the time we moved into our new home, rail movements in the yard were rare. I would have been at school at most relevant times, but I do recall once hearing a steam engine in close proximity and trotting down the road to witness an ex-LMS 2-6-2 tank shuffling over the road bridge, shunting a trail of battered wagons. I also recall seeing coal merchants' lorries entering and leaving the Depot and later, once or twice I saw, or heard, diesel shunters pottering about the yard. Knights Hill Depot closed in October 1968, less than three years after we arrived, and it was not long before the second road bridge, which had been integral to the yard, was demolished. British Railways sold the goods yard to the London Borough of Lambeth, who during the 1970s erected a housing estate on the site.

On the up side of the main line, just north of the original road bridge, was the signal box for the depot. It was a small timber affair, which was reached by steps that descended the embankment to a gate set in ancient railings by the roadside. The box closed less than a year after the depot, but its decaying structure remained for a while following its redundancy. To this day there remains a gate in the current metal fencing with steps leading up to the line side.

I left school in July 1970. The school was in Lambeth Road and was reached by catching the number 3 bus. I did not regularly use the local railways until I started work towards the end of 1971. There were odd excursions to the coast, which might begin at West Dulwich, or if I was going up to the West End I might catch a train to Victoria from the same station, or from Herne Hill. However, following the opening of the Victoria Line, catching a bus to Brixton and then picking up the Tube was usually the preferred option. I very rarely journeyed on the South London Line before my legal career commenced.

I continued to live in Dulwich until 1981, and so for ten years I commuted to Holborn Viaduct from Herne Hill. I watched the former exchange goods sidings that straddled the line to Loughborough Junction get overgrown with weeds and saplings, before their tracks were lifted and finally the land was fenced off prior to disposal for redevelopment. For a period on Fridays my firm closed at five rather than half five, so I would scurry from the office in Chancery Lane to Blackfriars to catch the 17.17 to Orpington, which started from one of the battered wooden terminus platforms that stretched out over the Thames. For some reason when I caught that service I always alighted at West Dulwich rather than Herne Hill. I am not sure why, because there was little difference in distance, or in the gradient to be ascended, as I struck out up towards home over the rising ground of Knights Hill. Occasionally I would find myself in the City at the end of a working day and could then catch a train from London Bridge to Tulse Hill, cruising above and beside the familiar heart of Dulwich before crossing Rosendale Road, catching a fleeting glimpse of home, before slipping into the gloom of Knights Hill Tunnel. It was a ten to twelve minute walk from the rear entrance of Tulse Hill Station, across Thurlow Park Road, before a steep climb up the eastern flank of Knights Hill was followed by a gentle descent along the broad curve of Lovelace Road.

People often talk about "culture shock": it is an accurate way to describe the impact of my family's move to West Dulwich. In Pullen's Flats every window overlooked a brick wall. We lived on the second floor and from the front room the nearest wall was forty feet away, on the opposite side of Penton Place, but the walls visible from our living room, kitchen and my parents' bedroom were often no more than twelve feet distant. I slept in the front room and at night, while there was little vehicular traffic given how few people in Walworth then owned cars, if I struggled for sleep there were a plethora of other familiar sounds. Shouts and ribald singing when two nearby pubs turned out, the bells of emergency vehicles speeding along Walworth Road, the fluty piercing pitch of a police officer's whistle, often followed by heavy footsteps and more distant shouts and whistles, as other constables joined the chase. An empty bottle knocked over, rolling in a gutter, perhaps smashing, or the hollow barking of dogs, one triggering a response from another until half a dozen canine voices choroused in an attempt to wake the weary, perhaps the dead. The railway line from the Elephant & Castle was to the east, so I rarely heard passing trains, but when the wind was blowing from the south-west the whistles of engines passing through Vauxhall could reach me.

Our house in Rosendale Road had gardens to the front and rear. It was set well back from the pavement where there was an avenue of tall, mature horse chestnut trees.

From my parent's bedroom they could gaze out over the "Village in the valley", to glimpse Honor Oak to the north. My bedroom was at the rear and I looked across the allotments to the top of Knights Hill, while to the right stood a copse of dense woodland, which sat directly above the northern portal of the tunnel. What was common to the view from all these windows was the sky. Suddenly we had access to light, space and perspective. At night I seldom heard cars passing to the front of the house. There was the occasional voice of a neighbour, the opening or closing of a door and the unfamiliar sounds of cats moaning, or a fox screeching. The foxes lived in the copse above the tunnel and during summer months I observed them patrolling the allotments as daylight faded towards dusk.

As well as the sports club, of which I was by then a member, I availed myself of other amenities offered by The Dulwich Estate. One of its tenants in the village was the Crown & Greyhound public house, a local institution. There was Dulwich Park and John Soanes's unique picture gallery, which stands in its own grounds close to the seventeenth century chapel and almshouses. Walks were taken in Dulwich Woods towards Sydenham, following the route of the by then departed railway line to Crystal Palace. Except when there were major events at The Palace, passenger numbers were never heavy and traffic was indeed suspended during both world wars. The line closed in 1954 and much of its remains were quickly interred. The station at Lordship Lane, despite its architectural splendour, was demolished to make way for a housing estate. The station house at Upper Sydenham survives, as do the two tunnels, but all that remains of Crystal Palace High Level is a tall retaining wall and a subway designed by Charles Barry Junior that led through to the Crystal Palace. The subway is Grade II listed, but not currently accessible by the public.

You can walk from Lordship Lane along Cox's Walk up through the woods where the old trackbed passes beneath a footbridge. Today you can continue along a footpath that follows the defunct route towards Sydenham, as far as the entrance to Crescent Wood Tunnel. Access to the tunnel is prohibited unless you are one of the many bats who roost within its depths. It is a pleasant stroll and one I made with several girlfriends, including the one who became my wife. The footbridge is still extant and is notable for being the location from which Camille Pissaro painted a down train leaving Lordship Lane Station. This was during his first period of exile in London in the early 1870s, when he resided in Upper Norwood. Today that painting hangs in the Courtauld Institute of Art.

The bridge carrying the former LCDR main line over Turney Road. The original span has long been replaced, but the lower ironwork remains. The crests visible either side of the pillar on the right are those of Dulwich College.

The Railways of Dulwich

From my bed I rarely heard the passage of trains along the Dover main line, which ran a quarter of a mile downhill and to the east, but I certainly heard the movements of trains along the South London Line. Down trains in particular could be heard for several seconds as they approached, clattered across the points that accessed the goods sidings, rumbled over the road bridge and then there was a further stutter as the wheels engaged with the second set of points connecting to the depot, immediately before the tunnel entrance. Once the final carriage penetrated the tunnel there was silence: no residuary echo or rumble. I do not recall the sounding of horns and London-bound trains would suddenly engage your attention at full volume, before fading once the road bridge was crossed. These regular navigations became familiar, particularly at night when they pierced the darkness outside my window and were oddly comforting.

I left Dulwich to marry and, in due course, raise a family. My parents remained in Rosendale Road for the rest of their lives. We finally sold the house in 2017 following my father's death. I had moved to Sutton, although our local station is Carshalton, and both those stations remain destinations further down the South London Line. My commuting from the early eighties until 2001, when I ceased to work in Central London, was at first into Victoria and for the final few years, London Bridge. I also frequented down Thameslink services from Farringdon, particularly once a half hourly service via Mitcham Junction was introduced. Gradually my ties to Dulwich loosened, but like any place you once called home, references to events in that former parish have a tendency to trigger reflection. One such instance was my purchase in the early 1990s of a Middleton Press publication, 'Mitcham Junction Lines', one of several volumes I acquired from a series authored by Vic Middleton and Keith Smith.

The attraction of this book was that it featured the South London Line and therefore covered both my former territory around Dulwich and also the route from Sutton through Hackbridge up to Streatham Junction, along which I passed daily as a commuter. The narrative comprised a series of photographs with captions portraying locations and features along the specified route. There were also extracts of ordnance survey maps and by way of a preface, a brief history. As I turned the pages, the journey began at Peckham Rye and proceeded southerly through East Dulwich and then North Dulwich. There followed an 1870 map showing the lines to the north of Tulse Hill, so detailing both the route through Knights Hill Tunnel and the LCDR link to Herne Hill; on the adjoining page was a larger scale replica of a map dating from 1916, which showed the layout of the old goods depot. When I turned that page there was an image of the depot's signal box and beneath, a photo of an electric unit buried in a wall of debris. The caption read:

West Dulwich Station stands beside the constantly busy Thurlow Park Road, part of the South Circular Road. The building is less ambitious in design than North Dulwich, resembling a villa rather than a palazzo.

"The 331 yard long Knights Hill Tunnel was the scene of an unusual accident on 27 December 1940 when an enemy bomb penetrated it and a local train ran into the debris. Sadly the driver lost his life in what was usually a safe place in wartime."

I was instantly affected by this image. It was many years since I last slept in my old bedroom and listened to the passage of trains running into and out of the tunnel, but I instantly imagined what had happened that bleak December night and how, when his train rattled over the final set of points linking the main line to the goods depot, the unsuspecting driver would surely have had no inclination disaster was only seconds away.

Over many years when from time to time I lifted this book from the shelf and flicked through its pages, I lingered longest at that image. A group of nine men of varying ages and rank are in attendance: one is purposefully positioned in what appears to be a guard's compartment; three are standing with enquiring expressions at the foot of the debris mound, while the others stand on the adjacent track. Those in the foreground only have their heads and shoulders visible. Most, if not all, appear to be railway staff, although some may have been members of a Rescue Party: those nearest the camera – two of whom are wearing trilbies – appear to be of higher rank than the men nearest the point of impact, one of whom is grasping a lamp. In the spring of 2019, several months after my retirement, a further casual review of this photo led me to search the web to see if I could locate any reference to this tragic incident, but I found nothing.

I persisted and having casually searched various volumes relating to railways in wartime without any reward, I decided to widen my enquiries and visited the websites for the London Metropolitan and National Archives. Two visits to the former led to me leafing through hundreds of incident reports filed by the Auxiliary Fire Service in relation to air raid damage sustained on 27 and 28 December 1940. I also looked at the Brigade's Incident Books for the months of December and January, but there was nothing relating to Knights Hill. Prior to that visit I had with the assistance of the Fire Brigade Museum obtained confirmation of the various principal fire stations in South London during the Blitz and of the numerous sub-stations, one of which I discovered was located at Rosendale School, literally across the road from the entrance to Knights Hill Goods Depot.

Walking from Farringdon Station through Clerkenwell on my way to the Archive I had been convinced there would be some reference to Knights Hill Tunnel and so it was deflating to find nothing. However, during the following days, I reasoned that notwithstanding the number of reported incidents for those two nights, some serious but the vast majority of a minor nature, I should continue my research. I had already decided to visit the local history archives maintained by the London Boroughs of Lambeth and Southwark. That was primarily to seek historical sources regarding Dulwich, its College and the development of the area in conjunction with the arrival of the railway. Rosendale Road and Knights Hill fall within Lambeth, whereas Dulwich Village and the bulk of the Dulwich Estate are in Southwark. The catalogue for the Lambeth Archive is posted online and disclosed that wartime records of the Lambeth Civil Defence Emergency Committee are available.

A short distance east of West Dulwich Station, the Dover main line crosses Alleyn Park with Charles Barry Junior's glorious Italianate confection, Dulwich College, in the background.

The Railways of Dulwich

As I strolled from Loughborough Junction Station on my way to the Minet Library, I recognised that the basis upon which I had originally conceived this essay about the railways in Dulwich and my family's interaction with the area was inexorably shifting. Somehow that photograph had become a catalyst for the entire enterprise; the fulcrum upon which my progress was now balanced. I realised I had formed an emotional connection with the event it portrayed and felt impelled to establish precisely what had happened. I had been cautioned that record keeping during wartime was haphazard and the press would not have been at liberty to report events widely. Adversity encountered by the populace in general might fuel propaganda disseminated by the enemy. Nonetheless, as I registered with the librarian and outlined why I wished to view certain documents, I recognised my project had become a quest and its original, broader scope had narrowed to what was a kind of tunnel vision.

The archivist could not have been more helpful, or indeed, more enthusiastic about my research. Having received an outline of my efforts to date he asked if there had been any fatalities. When I replied, "one", he referred me to a book compiled by John Hook: 'The Air Raids on the London Metropolitan Borough of Lambeth', which purports to contain details of each and every death registered in that district during the Second World War. While I reviewed that volume, he located the Civil Defence Committee Diary for the Lambeth Control Room. It did not take long to establish there were only two deaths recorded by John Hook for 27 and 28 December and neither occurred anywhere near Knights Hill.

My previous research had disclosed that Friday 27 December 1940 was the night the Blitz resumed after a lull of three days. There were almost no bomb damage incidents noted for Christmas Eve, Christmas Day or Boxing Day. The bombers had returned with a vengeance the following night. Lambeth's Civil Defence control room was in the basement of the Town Hall in Brixton. The Diary was a heavy leather bound ledger in which reports of incidents were recorded in long hand. When I reviewed the page for 27 December, I at last found confirmation of the event depicted in the photograph. There were three entries about Knights Hill Tunnel, one for the 27 and two the following day. They read:

"27/12/1940 – Knights Hill Rly tunnel Rosendale Road end. H.E. 1 Killed

28/12/1940 – Situation Report 06.00 hrs

Re: incident at Knights Hill Railway Tunnel: train driver killed and guard slightly injured. Damaged electric train removed. Total extent of damage to structure of tunnel not yet ascertainable, but roof of tunnel has been holed and both ends of line are completely blocked.

28/12/1940 – Situation Report 19.30 hrs

Further to our 06.00 hrs report for the incident at railway tunnel, Knights Hill: both lines are completely blocked and likely to be so for some time. Railway traffic between Herne Hill and Tulse Hill stopped owing to UXB confirmed in 16.14 hrs report."

So, the collapse of the tunnel roof had been caused by a high explosive bomb. This I believe was a relatively rare event in Dulwich: an article posted on The Dulwich Society website

Part of the track bed of the line to Crystal Palace High Level, as it passes through Dulwich Woods to the south of Lordship Lane and approaches the southern portal of Crescent Wood Tunnel. A group of volunteer conservationists loiter in the foreground.

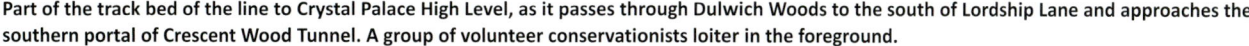

The aftermath of the collision on 27 December 1940 which resulted in William Burden's death.
British Rail

called 'The Wardens' Post' refers to a booklet published during the spring of 1946 by a former air raid warden, who had been based in an ARP hut in Burbage Road. He states that during the Blitz "more than twenty" high explosive bombs were dropped on Dulwich. My research suggests this particular strike was not the only bomb to drop in the immediate vicinity of Knights Hill. In the 'Herne Hill Heritage Trail', it is reported the then superintendent of the Peabody Trust estate in Rosendale Road was, together with his wife, killed during an air raid. The official bomb damage map shows the location of their cottage – which is marked as having been totally destroyed – but the map does not record a bomb impacting on open ground at Knights Hill.

Each incident recorded in the Diary had a reference number and the archivist duly produced a folder containing a sheaf of printed Message Forms on which someone in the control room had inserted details of the incident as the hours passed by. As with the fire brigade incident reports, handling these original, handwritten, documents was an extremely moving experience.

The first report does not have a time inserted. It simply states, "Train has run into obstruction". The location of the incident is not confirmed, and the section stating "Air Raid Damage" is not ringed.

The next Message Form is timed at 21.40 hrs. This time the incident is confirmed to be air raid damage:

"Position of occurrence: Tulse Hill… tunnel… midway.

Type of bomb: H.E.

Casualties: approx. 8

Whether trapped in wreckage: Yes

No fire, no blockage of roads.

Time of occurrence: 19.48

"Services already on spot or coming… 2 SP Carnac, 2 RP Salters Hill, 2 AMB 168"

Pilot engine called to withdraw train from tunnel."

At the foot of the Message Form is a statement: "NO FURTHER ACTION".

A separate untimed Message Form suggests the alarm was raised to West Dulwich Station and not Tulse Hill, or North Dulwich.

In the folder were several "Incident Slips" confirming instructions issued by the Civil Defence Control Room to emergency services. For instance, at 20.23 hrs the ambulance service was contacted and told "Enter Rosendale Road – bomb dropped through tunnel".

Other slips suggest instructions were sent to the Fire Brigade's sub-station at Carnac Street School in West Norwood and to another installation in Salters Hill, which appears to have been the base for rescue parties. It is recorded the fire brigade was asked to send two "pumps", but if they did respond why had there been no incident report among the records I previously inspected? The instruction issued to Salters Hill also advises, " to enter from Rosendale Road…" which clearly indicates, given the tunnel is reported to be blocked, the train had been running south from North Dulwich.

A further report was compiled at 21.47 hours and reads:

A. TULSE HILL railway tunnel mid-way H.E. 19.48 hrs

B. Bomb penetrated tunnel and trapped an electric train. Driver trapped in control room of train – dead – seven (7) passengers removed not injured.

C. R.P. endeavouring to remove driver. Pilot engine sent for, to withdraw train.

There are no further documents for that evening, but a few minutes into Saturday 28 December, a more detailed Supplementary Report was filed:

A. Tulse Hill railway tunnel mid-way H.E. 19.48 hrs

B. As 21.47 hrs supplementary, then add:

Train guard C. Vaughan, 45 Churchill Road, South Croydon, was slightly injured on forehead but has been able to proceed home.

Train driver Burden attached London Bridge dead.

Roof of tunnel holed. Both up and down lines completely blocked. Extent of damage to structure of tunnel not yet possible to ascertain.

C. Two pilot engines arrived at Tulse Hill Station at 23.00 hrs to effect removal of damaged electric train. Two rescue parties and one S.P. standing by to co-operate with railway services.

D. Nil

E. The following passengers rendered assistance to train guard:

D. Skinner, 1, Tolcroft Road SW16 attached LNER Bishopsgate Goods Depot.

C. E. Leather (RAF), Kent Lodge, Chatsworth Way SE27.

So, I now had a surname for the driver and knew slightly more about the lightly injured guard. How, I wondered, did Mr Vaughan make his way home in the blackout to South Norwood and what was his state of mind, given the trauma of the collision and the loss of his colleague?

But what leapt out at me from that flimsy pink and white page was why had the locomotives been sent to Tulse Hill Station? The various slips clearly indicate the damaged train had entered the tunnel from the north, the driver was trapped in the compacted cab and both lines were blocked. So, if the rescue engines were sent to Tulse Hill they had no immediate means of providing assistance. The shortest route to the north end of the tunnel would have been via Herne Hill to Loughborough Junction, before reversing onto the line to Catford through Denmark Hill where, having reached Peckham Rye, they could reverse again and run south towards Dulwich. However, at some point that night the line from Herne Hill to Tulse Hill was closed, because two unexploded bombs were discovered behind 275 Norwood Road, directly adjacent to the railway viaduct.

Despite my assumption it was very unlikely any reference to this incident would appear in the press, I was encouraged by the archivist to check the January issues of the 'South London Press'. His hunch proved correct and on the front page of the edition published on 4 January 1941 the following short item appeared:

TRAIN BURIED IN TUNNEL

An electric train had just entered a tunnel in a London suburb when it was buried by a subsidence.

The driver, Mr William John Burden (45) of 127, Shardeloes Road, Brockley, was killed.

The funeral will take place tomorrow (Saturday) at Nunhead Cemetery.

The train, which was travelling from London Bridge, was seriously damaged in the forward section, but there were no casualties because the only passengers were in the rear of the train.

Breakdown gangs were seen at work to repair the damage and there was little traffic delay.

This brief report appears to corroborate previous evidence that motorman Burden's train was running on the down line from London Bridge towards Tulse Hill.

I now knew the full name of the ill-fated driver, his home address and details of his funeral. I wonder how many of the paper's readers suspected this story did not disclose the whole truth and that William Burden had met his death as a consequence of enemy action. Historically there were grounds for suspecting the collapse of this tunnel could have been caused by landslip, or subsidence. As recounted earlier, the soil at Knights Hill is clay and back in the sixteenth century Henry Knight had excavated clay to manufacture tiles. The authors of 'Herne Hill Heritage Trail' report that clay extracted when the railway cuttings and tunnel were dug in the 1860s was deposited on the hill and as a consequence the ground became very unstable. A major landslip occurred in 1919 and it is suggested that was why most of Knights Hill was left undeveloped, with much of the land turned into allotments. That remained the case until the 1970s when Peabody Trust built a further housing estate that cascades from the summit down the western slope, overlooking the approaches to Tulse Hill Station.

Despite the caption to that photograph, my research established the nature of the incident at Knights Hill may not have been quite so "unusual". While browsing in a bookshop I found a reference to similar damage having being sustained a mile or two east of Knights Hill, when during the Blitz a high explosive bomb dropped on open land at Sydenham Hill and caused the roof of the tunnel below to collapse. On this occasion it seems the blockage was discovered before a train could run into the spoil, but apparently the Dover main line was closed for approximately a week.

Having discovered precisely what happened that fateful evening and having identified the train crew, my thoughts turned to what the records of the Southern Railway Company might further disclose. So I visited the National Archive at Kew to peruse the board minutes of the Company during the period of the Second World War. Initially, I was optimistic there would be some reference to this incident and the disruption it must have caused. However, as I began to read through the minutes of the monthly meetings for the latter part of 1940 I realised there were few specific references to interruptions to services caused by enemy action. Each month the General Manager updated the directors regarding a range of departments including: Traffic, Engineering and Estates. As the Blitz began there were increasing mentions of agreements to abate rents where lettings of railway property were frustrated by damage to, or destruction of, premises and there were also regular reports of bomb damage to the Grosvenor and Charing Cross hotels and of the need to carry out consequential repairs to those assets. There was also the regular noting of agreements with the Government and armed forces for the creation of additional sidings, crossovers and running loops, but the negative consequences of the Blitz on the railway's activities were notably absent.

So, as I turned the wide pages of the two bulky minute books, my sense of anticipation rapidly diminished. By the time I reached the conclusion of the minutes of the final board meeting held in 1940 I had though noted a new regular item, which had begun to appear in the General Manager's Reports and became of increasing interest, as I felt sure at some point there would be a mention of William Burden. Each month the Company minuted the award of special gratuities paid to the widows of employees who had been killed by enemy action, either while on duty, or in a few cases, when not at work. The typical amount awarded during 1940 to 1942 was £15.00, or on a few instances, £20.00. Later in the War gratuities of £25.00 were paid.

As I worked my way through the minutes for the first half of 1941 I was convinced I would find an award to the widow of William Burden, but regrettably I did not. Perhaps, I reflected later, he was not married? However a subsequent review of the 1939 Register disclosed William Burden was living at 127 Shardeloes Road, Brockley, with a woman I assumed was his wife; Florence Burden (nee Whitmore) and there were two teenage children, Betty and Guy. If that was so, why did the Company apparently not grant an award and recognise William's sacrifice and his family's loss? Perhaps not all such awards were minuted, but given the number who were shown to have benefited and the variety of their grades and trades, I fear that was unlikely. It has left me with a sense of grievance.

As a former lawyer, I was not deflected by the dearth of relevant information and continued to leaf through the minute books in the hope of finding something pertinent. By the time I reached the minutes for the board meeting held on 26 November 1941 my diligence was to a small degree rewarded. By this time each General Manager's Report contained an item detailing works that had either been tendered, or completed and paid for by the government. I assume this was compensation for the cost of repairing damage to the railway infrastructure and for reinstatement of services. So, the November minutes contained the following item:

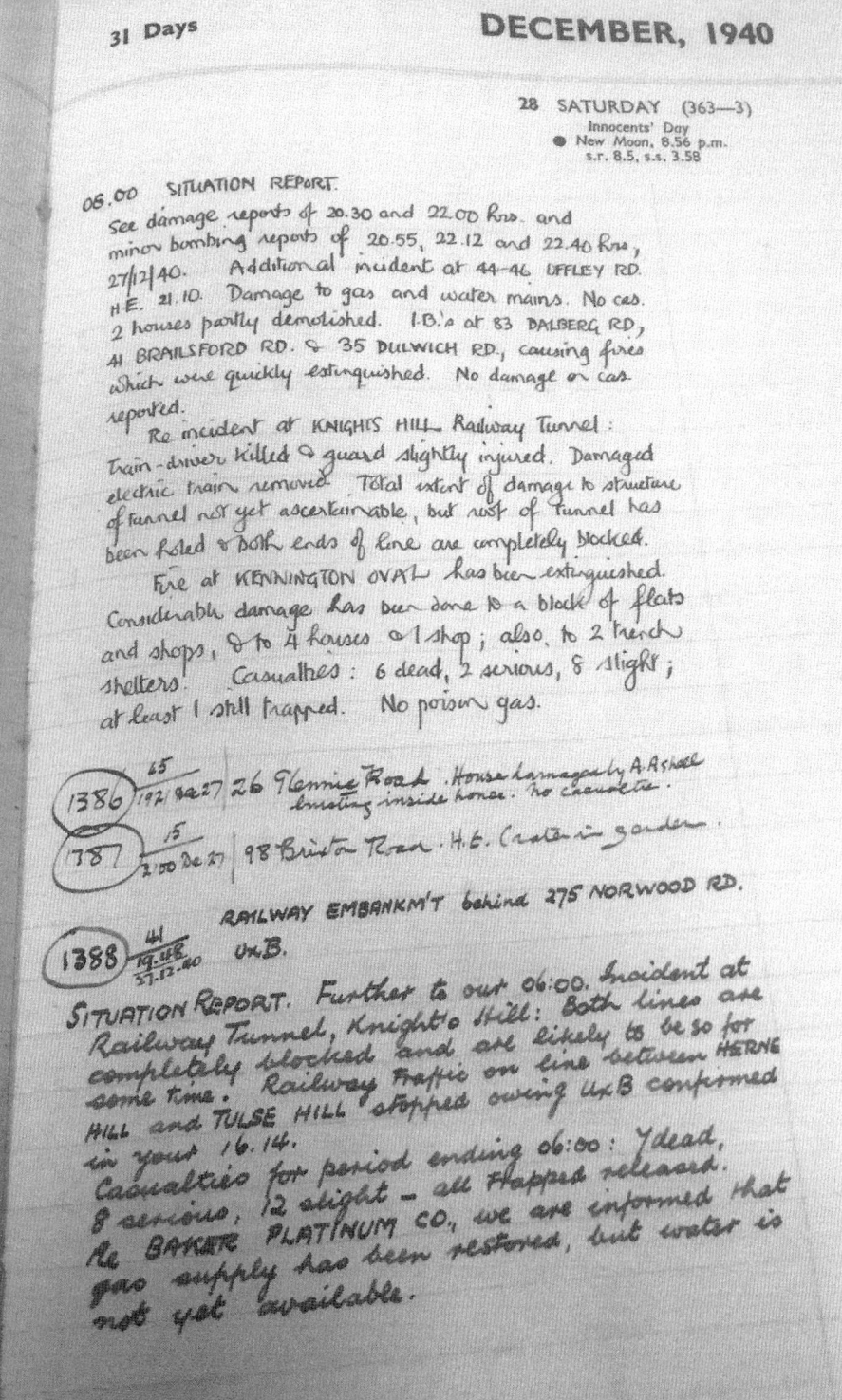

Lambeth's Civil Defence Emergency Committee Diary, showing the entries for 28 December 1940: 'Innocents Day'.

"Works completed under prime cost contract issued by the Minister of Home Security Limited:
  "Penge Tunnel"
  Excavation Work
  R. Robinson & Company Ltd
  £282. 7s. 4d."

Could this work refer to the bomb damage I had read about?

Then in the minutes of the board meeting held on 30 April 1942, the following item appears:
  "Works completed under prime cost contract issued by the Minister of Home Security Limited:
  Tulse Hill – North Dulwich
  R. Robinson & Company Ltd
  £3,203. 7s. 1d."

There is no specific mention of Knights Hill Tunnel, but surely, given the nature of the damage sustained in December 1940, did the "Works" include clearing the debris and rebuilding the bore of the tunnel? I suspect, as is often still the case, the contractor completed its task long before the cost was reimbursed by central government.

William Burden and his guard were but two of thousands of railwaymen who daily risked their lives, carrying out their duties during an ever more distant period of conflict. A great many died. The bombs that fell on Knights Hill and the Peabody Trust cottage were not the only bombs to fall in that vicinity. During the summer of 1944 a pair of semi-detached houses which stood at the point where Lovelace Road sweeps northward and joins Rosendale Road were destroyed by a V1 flying bomb. Four people, including a sixteen year old Alleyn's schoolboy, were killed. Two months later another flying bomb destroyed houses further up Lovelace Road. Latterly, a plaque was erected by The Dulwich Society to commemorate those tragedies and you can find details of them and other nearby incidents online. However, with the exception of the short article in the 'South London Press' and the photograph in 'Mitcham Junction Lines' (where his name is not mentioned), the circumstances of the death of William Burden are not recorded. His death certificate states his body was found on the railway line at Tulse Hill and his cause of death was "due to war operations".

The coming of the railway to Dulwich brought great benefits to that area and to the Dulwich Estate. Similarly, moving to West Dulwich greatly enhanced the long lives of my parents and the life experiences of their son. We moved from the grey grime and harshness of Walworth to the light, greenness and relative solitude that characterised our new home. Allotments on the hill behind and playing fields to the front; the broad carriageway of Rosendale Road with its sentinel avenue of horse chestnut trees that blossom each Spring and rain conkers come Autumn. Once, Henry Knight would have gazed from the hill that bore his name towards the towers and spires of Tudor London. Today people dig their allotment plots and when they pause from their labour they can peer northwards towards Canary Wharf, the Shard and those other shiny towers of twenty-first century London, which twinkle red and silver like Christmas trees as dusk descends. In the valley the trains – these days primarily carrying commuters – scurry from east to west and north to south crossing bridges, along viaducts and embankments and through the tunnel designed by Charles Barry Junior, all constructed over one hundred and fifty years ago. Charles practiced his profession at the direction of and for the benefit of Alleyn's College of God's Gift, Dulwich: a village that was then, and in many ways remains, a place apart.

Bibliography:

Norwood and Dulwich by Allan M. Galer
Planning a London Suburb 1882 -1920 by Bernard Nurse
Victorian Suburb (a study of the growth of Camberwell)
    by H. J. Dyos
History of Dulwich College by William Young
London, Brighton and South Coast Railway
    by John Howard Turner
The Air Raids on the London Metropolitan Borough of Lambeth
The Wardens' Post edited by George Brown
The Foreign Goods Depots of South London by Edwin Course,
    published in *The Railway Magazine*, November 1960

Acknowledgements

The London Fire Brigade Museum

The staff at the Local History Archives of the London Boroughs of Lambeth and Southwark

The S. C. Townroe Archive – in Colour
Part 4

One of the strengths of the SCT archive is the endless variety that exists: trains, engines, locations, derailments of course, and with each sub-divided even further.

For this issue's selection we have deliberately tried to include literally a bit of everything and starting with bits of engine that really should not have broken. Tantalisingly we have the usual date and some detail but not the full circumstances so as before, if anyone can fill in the gaps we would be grateful.

We have also managed to encompass all three sections of the Southern although looking through for the future, what is missing is anything from the far west.

However, hopefully that will not put you off for next time when we hope to feature: Bricklayers Arms, No. 60008 being shipped, various 0-6-0 types, inside Eastleigh Works, and on the Alton line.

Above and opposite page: The failure of the crank axle of No. 35020 at Crewkerne in 1953 is well documented (SCT does not record any views of this), but two years later on 16 November 1955 he photographed the result of failure of the coupling rod of sister engine No. 35016 that occurred at Gillingham. The views illustrate the scene trackside and also after the parts had been removed back to Eastleigh works with consequential damage found to the right hand leading axle box. Repairs were effected and the engine returned to service. Bradley appears not to mention the event but in 'The Book of…' there is mention that "…fractures occurred on all four coupling rods and cracks appeared in the axleboxes near Gillingham……The result was a redesign of the coupling rods and steel axleboxes on the rebuilds, new type crankpins". No. 35016 did not return to service until after 31 December that year. There is no mention of a similar failure on No. 35015.

The S. C. Townroe Archive – in Colour

Above and opposite top: **Coincidence perhaps, but these three views purport to show axle box and component damage to sister engine No. 35015 but on the same date. So was it a case of two engines with similar failures or a mere slip of the pen with all relating just to No. 35016?**

The S. C. Townroe Archive – in Colour

A simple cab view of an unidentified Merchant Navy at Southampton in 1951 – but then how many colour cab interiors exist from 70 years ago…?

Altercation at Eastleigh on 9 June 1964 and not the fault of either driver. Instead the derailment was caused by the point blades not being across correctly and in consequence not locked for the move. SCT does not record if this was a signalman's or track error/defect.

As we know Stephen Townroe was a regular attendee at derailments, both within his own area and when he was on call for other districts. One of those 'out of area' jobs is shown here at Strawberry Hill in June 1953. The motor coach of unit 4318 has somehow managed to get itself slewed over several roads at the same time, no doubt with consequential disruption to the working of the depot.

The S. C. Townroe Archive – in Colour

Another 'Olde England' job was at Eastleigh with a Mogul deposited in the dirt at Jubilee sidings, Eastleigh, on 19 November 1962. The Eastleigh – Portsmouth line runs to the right.

Overleaf: To attend to the chimneys of the Millwrights' department at Eastleigh Works, the Eastleigh crane was capable of having its jib extended. It is seen here with the extension in place in February 1954 and then five years later having the attachment put in place. When working in this extended mode the crane could not be called upon for emergency or other work and cover was instead provided usually by the crane based at Salisbury.

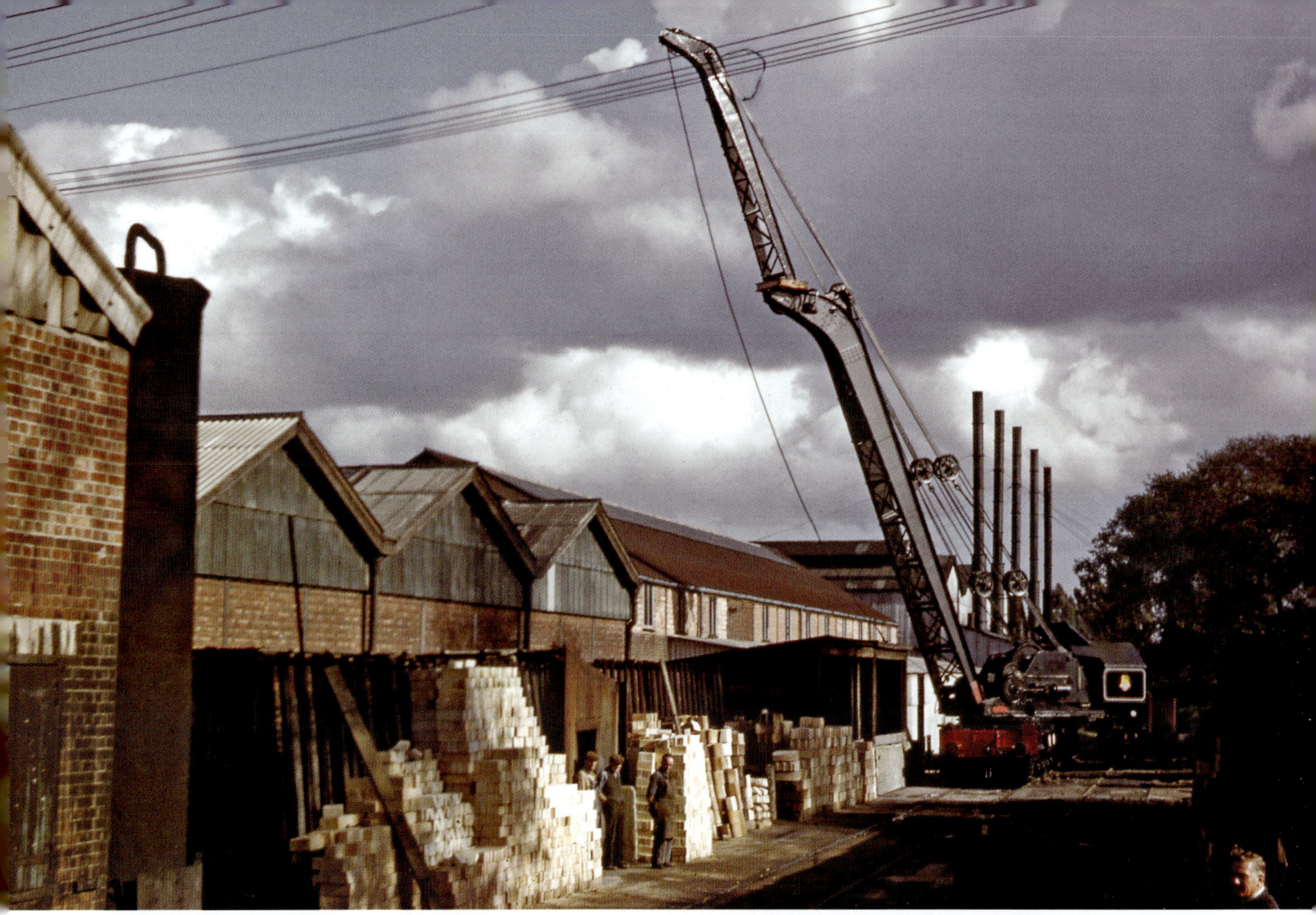

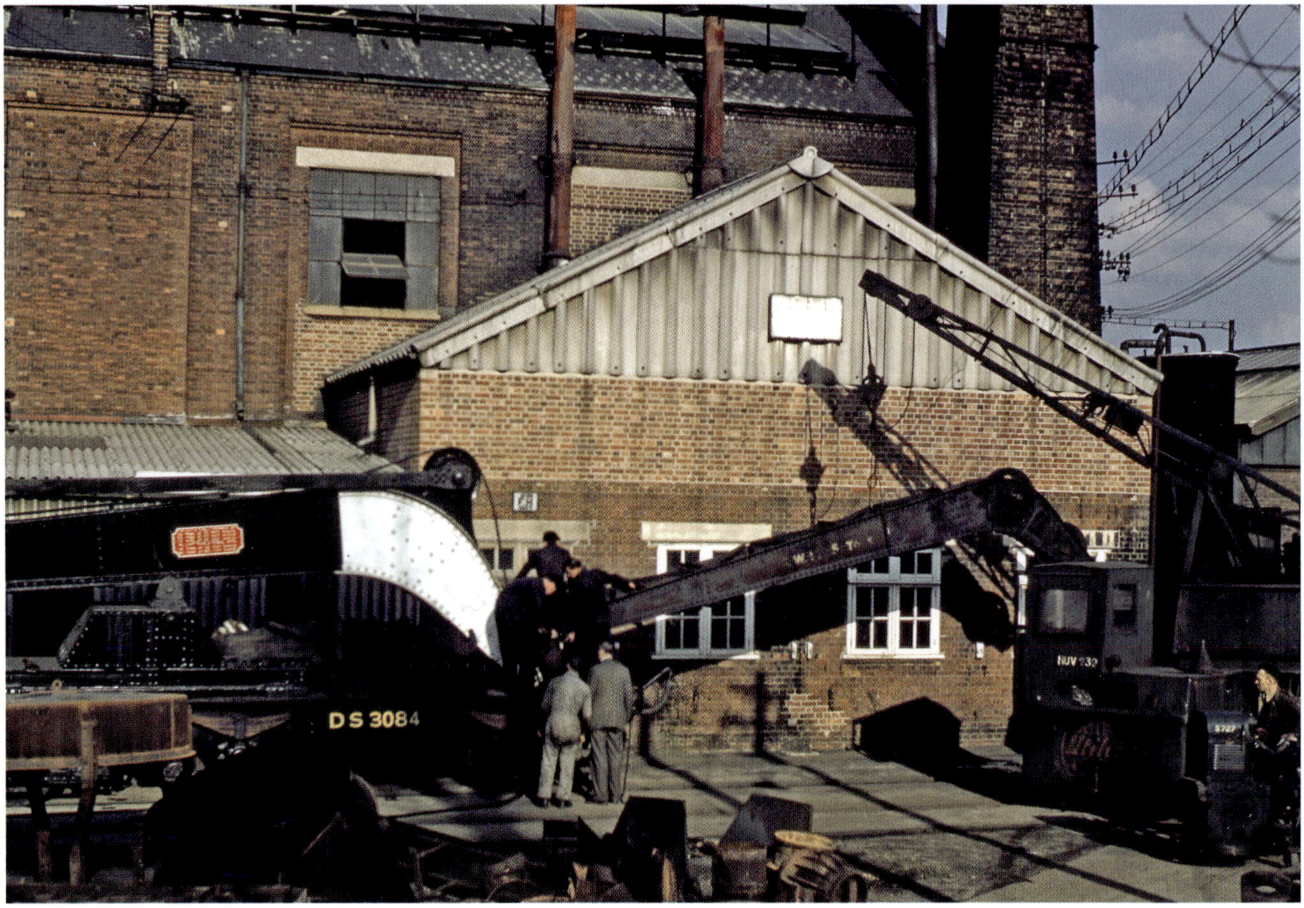
DS 3084

The S. C. Townroe Archive – in Colour

The fortnightly examination of Bulleid diesel shunter No. 15233 (notice the Bulleid wheels) outside Eastleigh shed in May 1953. Whether the open air work was by choice is not reported.

Pull-push working at Paddock Wood in 1956, the presence of the freshly painted vacuum-fitted van on the opposite end will be noted. Might this be a Hawkhurst branch train?

41

Maidstone-bound service leaving Paddock Wood in 1956. The three-coach set is slightly unusual in having a 'Birdcage' at one end only.
Still in 1956 we have M7 No. 30035 near Corfe Castle and seemingly propelling in the direction of Swanage.

Same branch line and this time clearly at Corfe. In the goods siding is '700' No. 30695 soon to be passed by a Mogul on its way to the terminus.

Pastoral scene at Kimbridge Junction not far from Romsey. Crossing the River Test in 1958 is a T9 on an Andover train.

Back on the Eastern section but this time at Tonbridge where 'H' class No. 31544 is taking water. The working/train is not reported although from the single disc it could well be a Maidstone West train.

Having seen pull-push working on the Eastern and Western sections we conclude with one on the Central lines – near to Bramber. SCT pressed the shutter seemingly at just the wrong time to capture the engine number, but no matter, it is still a scene impossible to recapture today.

Down To Earth Part 4
Ex-SECR Stock
Mike King

We will now turn our attention to the South Eastern – where probably even more variety may be found, thanks to the fact that prior to 1899 there were the two separate companies. Groundings of both LCDR and SER stock took place, although from what date is perhaps more difficult to determine. Certainly by 1920 vehicles were being sold off (to Dungeness amongst other locations), while lineside use of vehicles started well before 1900. What may be stated is that very few modern (ie. post-1910) SECR 54ft and 60ft 'birdcage' coaches became grounded and, with one notable exception that will be mentioned later, no corridor coaches were ever used in this manner. Not that there were ever very many of them in the first instance! Plenty of earlier (50ft and under) 'birdcage' vehicles were grounded – probably because the majority of these were withdrawn in the years before 1950.

Ex-LCDR 25ft four-wheeled brake third No. 318, seen grounded at Sevenoaks Tubs Hill goods yard on 16 March 1952. This was allocated LCDR No. 752 in the 1897 renumbering scheme but never received it: instead becoming SECR No. 3272 and ran at one end of 11-coach suburban 'block' set no. 1 – at least by 1905. There were quite a few of these sets, numbered from 1 to 37, and some were up to 14 coaches long, used for suburban services out of Victoria, Holborn Viaduct and Blackfriars, plus some shorter sets for various Chatham section branch line trains. The stock was generally withdrawn between 1913 and 1920, but a few rakes (or halves of rakes) survived into Southern Railway ownership to be allocated SR duplicate set numbers 01-018 (no relation to their former SECR set numbers), running until about 1926. Coach 3272 was from a batch of ten dating from 1880, was withdrawn in 1914, reinstated and finally taken out of service four years later. Presumably, it was grounded at Sevenoaks soon after. It is clearly now derelict – one of the missing doors may be seen (upside down) against the brake end, while the final vestige of a later SECR set number appears on the coach end. *D. Cullum*

45

The late Roger Kidner recorded, in his 1974 Oakwood Press book 'Southern Railway Rolling Stock', that the SECR was probably one of the greatest users of grounded bodies. He put this down to two factors – namely the permanently impoverished state of the railway and that fortunately their early (mostly contractor-built) carriage bodies were exceedingly robust! He goes on to record a few examples, including what may have been a very ancient SER 4-compartment first (with central luggage compartment) doing duty as a coal order office at Farningham Road station, seen by him in the 1930s. It had upswept panels at the ends rather in the style of the old road stagecoaches and dated from as early as 1850. He also noted very early ex-SER bodies at both Maze Hill and Reading – seen and photographed by him during the 1930-33 period, while former LCDR four-wheelers were also commonly seen at that date. Withdrawal of the latter began soon after the operational combination of the two companies (they never actually amalgamated) of 1899 – although a few survived in passenger traffic until as late as 1929. The Bluebell Railway has restored at least one such vehicle – their 26ft brake second, LCDR No. 114, withdrawn in 1926 and recovered from Bosham in July 1977 – although it took until November 2006 for it to be returned to passenger traffic, mounted on a shortened utility van underframe.

Although the SER/SECR bequeathed a large number of both four and six-wheelers to the Southern, relatively few of these were grounded after about 1925 – simply because many were re-used by the Southern in electric stock conversions. This might sound surprising, but many bodies (including most of the surviving four-wheelers) were relatively new (the last were built in 1901) so, in time-honoured Sir Herbert Walker fashion, economy played a considerable part in the 'new' electric stock. This might have reduced initial capital expenditure – something Sir Herbert was always keen to avoid – but perhaps stored up ongoing maintenance problems for the future.

A slightly more modern 26ft LCDR brake third, with side duckets instead of a roof observatory, is seen at Chart Road brickworks, west of Ashford, on 24 November 1950. This served as an office and was clearly connected to the telephone system, complete with external bell to alert the staff working outside to the fact that a call was coming in. No mobile telephones in those days!! The identity of this coach is unknown, but 1889-vintage brake second LCDR No. 114 as restored on the Bluebell Railway is structurally very similar. That coach lasted until 1926 in SR duplicate set 02, by then numbered as SECR 3068, and was grounded at Bosham until 1977. *D. Cullum*

When discussing grounded carriage bodies, this picture probably says it all! Three former LCDR 28ft 5-compartment thirds have been prepared for their new roles and are seen awaiting road transportation to their new owners, in Ryde St. Johns yard, possibly in early 1934. Neither the date nor the identities of the vehicles are known for certain, but as Alistair Macleod's tenure on the Isle of Wight spanned 1928 to April 1934 and as three such coaches, SR Nos. 2510, 2512 and 2513, were withdrawn together in February 1934, it is suggested that these are the coaches pictured. Furthermore, coach 2513 is known to have been grounded as a henhouse at a farm near Smallbrook Junction – only about a mile away – while another ex-LCDR third went to a site just a little further north. Clearly, mounted on carts as these bodies are, they are unlikely to be safely taken far by road in this manner. It may be seen that just the lettered panels have been repainted – to obliterate the carriage numbers, ownership and class designations on the doors, so presumably they remained in lined olive green livery for their new owners – at least initially. *A. B. Macleod*

From 1930, transfers to the Isle of Wight included ex-LCDR bogie coaches and a total of 41 were sent between then and mid-1934. This is SR Diagram 144 (later 144A on removal of side lookouts) 45ft 5-compartment brake third No. 4115, seen at Haven Street in July 1976. This had been built by Brown Marshalls in 1898 as LCDR No. 1185. It later became SECR No. 3411 and SR mainland number 3247 – the latter renumbering taking place at Ashford in April 1924. Transfer to the Island occurred in June 1933, when the coach assumed the identity of 4115. Until that time the body was still fully timber-panelled like the coaches seen in the previous pictures, but was subsequently steel-sheeted. The coach ran in 4-set 496 – based at Newport and this number is still visible on the brake end. Withdrawal came in May 1948 and traces of its 1945 repaint into malachite green are still visible. It was then grounded at a farm near Atherfield, on the south-west coast of the Island, and was recovered by the Isle of Wight Steam Railway in 1975 – one of the first coach bodies to be acquired by them. Unlike most other ex-LCDR bogie coaches grounded on the Island, this example remained in one piece – many were split in half before sale. To date little restoration has been carried out, but it is hoped that before long it will be mounted on a suitably shortened SR scenery van underframe and restoration to passenger service will commence. *R. Newman*

Spot the railway carriage! The Thames barge *Horselea* is seen moored alongside the Metropolitan River Police Waterloo Pier adjacent to the Embankment in Central London on 14 March 1950. Close examination reveals that the superstructure is actually an ex-LCDR 45ft 7-compartment composite coach. There were 14 of these and they became SR Diagrams 296 or 297, depending on whether they were built as bi-composite (1st/2nd class) or tri-composite (1st, 2nd and 3rd class) vehicles. There were slight differences in compartment dimensions between the two designs. Ten of the coaches were sent to the Isle of Wight and several returned to the mainland for scrapping in 1948/49 and it is thought that this was one of them. The four that stayed on the mainland were all withdrawn in 1930/31. So here we have, not so much a grounded body – more a floating body! There is clearly a use – and a prototype – for everything. The tram on the Embankment is also worthy of comment – would a knowledgeable reader like to fill in the details? *K. G. Carr*

Several outlying suburban stations had carriage sidings; and staff accommodation at many of these was provided using old carriage bodies. Blackheath had a great collection of ancient ex-LCDR coaches, but Grove Park held two ex-SER 29ft 4-compartment composites. The photographer recorded both the coaches and the cleaning staff on 30 April 1948. Denis Cullum also thoughtfully measured up one of the carriages and this enabled the writer to make a drawing of it – his drawing is dated 24 October 1994, so it took an awful long time to make use of Denis's sketch! Thirty of these six-wheeled vehicles were completed by Cravens in 1888/89, numbered as SER 2245-74. The two end compartments were second class (later third), the two in the middle were firsts. Most were altered over the years – some being reduced to four wheels, many had close-couplings and short buffers fitted, while one was uprated to all-first and one other demoted to all-third. All received Southern numbers and it is thought that the Grove Park bodies were the electrically-lit pair 5776/83 (most retained their gas lighting to the end), withdrawn on 1 December 1934. The Southern diagrams were 503 for the first class vehicle, 85 for the third and 380 for those composites that lasted long enough to be allocated a diagram number. Two more composites (SR Nos. 5788/89) were grounded at Selsey Beach in 1932 and, needless to say, Denis found these on his travels as well! *D. Cullum*

Down to Earth Part 4

A slightly older SER carriage with an arc, instead of a semi-elliptical shaped roof profile, grounded on a farm at Appledore (Kent) and photographed on 21 June 1948. This was also built by Cravens of Sheffield, in July 1887, and was SER No. 2134 – a 31ft 4-compartment brake third. It became SR No. 3615 and ran in SR excursion sets 715 and later 724 until withdrawn in December 1935. There were ten such coaches built by Cravens but the SER completed a further batch of very similar vehicles in 1888/89. Despite several of them lasting until 1935/36, they do not appear to have been allocated a Southern Railway diagram number. *J. L. Smith*

Bacon sandwich and coffee anyone? A feature of the centre of Lancing village from 1935 until into at least the mid-1950s was 'The Lancing Snack Bar' - situated just opposite Lancing station and the Luxor Cinema so it has clearly not moved far away from Lancing Works. Whilst undoubtedly popular with the locals, it was also an interesting (and unique) railway carriage. Formerly SER 33ft saloon brake first SR No. 7765, it began life in August 1886 as SER inspection saloon 2124 and for this purpose had a small end compartment containing two 'bucket' armchair seats and large end windows overlooking the track. This is at the left-hand end in the broadside picture – the sign 'Snack Bar Open' just visible above the period car occupies the upper portion of one of these windows. The coach featured in 'The Engineer' magazine for 21 August 1891, where a full description of the interior fittings (in the usual flowery language of the period) was given. Also incorporated was a comfortable saloon (the portion with the two larger side windows), a lavatory, a first-class compartment and a guard's compartment (right-hand end in the broadside view), allowing the coach to function as a complete train – usually propelled for inspection purposes by the locomotive. As such, the vehicle was used by the railway's 'top brass' for trips out (jollies?) and came to Southern Railway ownership still so utilised. However, in 1926 it was transferred to ordinary passenger service and renumbered as 7765, running at one end of excursion set 719 until withdrawn in May 1935. SR Diagram 547 was allocated. The broadside picture was taken about 1946/47, the ¾ view on 10 June 1950. *The Lens of Sutton Association & D. Cullum*

Surprisingly, the SER/SECR continued to build four-wheelers for suburban traffic until as recently as 1901, and all survived to become Southern Railway property in 1923. Starting in 1894, a number of 14-coach close-coupled sets were completed and these were followed by further batches of 7, 8 and 15-coach sets. All 432 coaches were 27ft long but were otherwise shorter versions of the then current six-wheelers and bogie stock – so they presented a surprisingly modern appearance despite their small size. A total of 75 received SR carriage and van stock numbers and lasted to 1929/30, just 13 were scrapped in 1926 while all the rest were incorporated into electric stock on new 62ft underframes between 1925 and 1928 – two four wheelers to each new electric vehicle – without having SR steam stock numbers allocated. This former 5-compartment third is one of those withdrawn in either 1926 or 1929, but beyond that no further identification is possible. It was grounded at Rye Harbour as a dwelling and was photographed on 21 September 1950. The writer saw it there, both in 1995 and in February 2010, still inhabited so it might even be there today. *R. C. Riley/The Transport Treasury*

At first glance this coach body seen at New Cross Gate yard on 26 April 1947 posed something of a puzzle but on closer examination it turns out to be two bodies joined end-to-end. Well, not quite, as the larger right-hand portion (with the paler roof covering) is to SR Diagram 48; a 48ft lavatory third – ex-SR No. 943, while the far end two compartments with a lavatory between them come from another ex-SECR coach – whose number is unrecorded but was probably another lavatory third or a composite. This was a not unknown situation – as well as bodies being split in half and placed at two different locations, so when searching out grounded bodies one had to be ready for anything! Built as SECR No. 867 by Gloucester RCW Co. in 1906, the coach later ran in SR 'long' set 909 on Kent Coast excursion trains, then on troop trains until damaged by a V2 rocket attack at New Cross Gate in October 1944. There were 19 examples of the type, plus five more with an identical layout but two feet longer – allocated SR Diagram 49. Several were loaned to the War Department towards the end of World War 2 and the late Roger Kidner recalled travelling across Belgium in 1945 in a train of these SECR vehicles – numbered as WD set 5. Several of the 48ft examples were grounded and the writer remembers measuring one up at Ashford Wagon Works as recently as 1974 – by which time both it and the works were practically derelict. Others were seen grounded at Hither Green yard and at Dunton Green station. The LNER clerestory brake coach seen behind is a mystery. *R. C. Riley/The Transport Treasury*

At Ashford Works on 20 April 1949 we see 45ft 3-compartment 'birdcage' brake third No. 3255 – the number may just be seen in the second eaves panel from the right. This was built by Metropolitan as SECR No. 2300 in July 1900 and was originally destined for boat train service – hence the large luggage van space that occupied over half of the vehicle. Co-incidentally, it also ran in SR 'long' set 909 and was grounded at Ashford in September 1944. The last example in traffic was SR No. 3266; withdrawn from 'long' set 696 in December 1956 but several others survived a little longer in departmental stock. *J. H. Aston*

A 51ft 1in composite, seen grounded at Horsham goods yard on 28 April 1949. There were a number of 50ft and 51ft SECR composite coach designs and many were only built in small numbers. This one, SR No. 5394 to Diagram 312, is particularly interesting as it was one of just two examples. Built in July 1905 for boat train service as SECR No. 3805, it had the layout: 3-Lav-3-1-Lav-1-3-3-Lav – the lavatory right across the far end of the coach being a particularly unusual feature – although not quite unique in a compartment coach. It later ran in trio-B set 526 between two 'birdcage' brake coaches and was withdrawn in September 1942. Its partner, composite 5393, was also withdrawn in 1942 but this became a departmental coach, renumbered as 1705s. *D. Cullum*

This 44ft 6-compartment composite coach was an SER design that spanned over into SECR days – and some 35 were built between March 1899 and January 1901. Originally they had a second class compartment at each end, with four firsts in the middle – the seconds being re-designated third on the formation of the Southern Railway in 1923. As such they became SR Diagram 292, with running numbers 5185-5219. With the exception of coach 5194 (withdrawn in July 1933) all the others were reduced to third class between 1935 and 1937, being renumbered as SR 858-891 and re-allocated to Diagram 57. In this form many were used as loose strengthening vehicles – several in the west of England – until withdrawn between 1938 and 1951, except for No. 871 which contrived to last until July 1955. Quite a number were either taken for departmental service or grounded after withdrawal. This example was SECR No. 2367 and was built in November 1900, later SR composite 5208 and all-third 880 from December 1935 until grounded at Shalford East sidings in February 1943. It was photographed on 21 April 1951. Shalford East sidings were put in during World War 2 for the build-up to invasion and comprised six through roads and four dead-end sidings – nicely out of the way but well-served by the Redhill-Guildford-Reading cross-country route. They were later used intermittently for wagon storage, for pre-assembly of rail sections during the Bournemouth electrification and finally to store withdrawn electric stock pending disposal. The writer recalls seeing the sidings derelict and completely empty in August 1964 under a carpet of weeds, but only a year or so later were in full use as an overspill depot by the permanent way department from Woking Yard. Once this activity ceased, he again visited during 1971-73 to measure up the various electric stock coaches stored therein. At least this was safely away from the live rail! By entering from East Shalford Lane end, few BR staff would be encountered! *D. Cullum*

Not so far away, at Farley Green, to the south of Gomshall, a Ministry of Supply depot was established during World War 2 – one of a considerable number situated at remote locations around the country. This one utilised four carriage bodies – an ex-LSWR 4w luggage van and three SECR bogie vehicles. Once the war was over, this was converted into a holiday camp under the name of 'Treetops'. Today we might look askance at the notion of spending a country holiday in an old railway carriage, but after six long years of hostilities, people were only too glad of the opportunity to relax – anywhere. Two of the coaches were former composites but this one was an ex-saloon – SR Diagram 616 50ft coach No. 7916 – grounded at the site in September 1942 and photographed on 2 June 1948. It was built by the Midland Railway Carriage & Wagon Company in 1904, as SECR No. 3512 – one of a pair – and consisted of a lavatory at each end of the coach, with two large first class saloons in the centre. There were several similar third class saloon coaches (to SR Diagram 621) and these were also used for office accommodation (either ON wheels or grounded) at various locations during and just after the war. Just how long the holiday camp survived is not known (it was still operating in 1953 but derelict by 1959) but just in case anyone is moved to visit, the site is today occupied by a large private house with no trace of any railway interest. *D. Cullum*

Several ex-LCDR van bodies were also grounded. This is a 28ft former guard's van seen far from home at Exmouth Junction carriage & wagon workshops in the mid-1950s. It could have been grounded there at any time between the mid-1930s and the early 1950s. If grounded at the earlier date it would have come straight out of ordinary traffic, but if grounded later might have served as a departmental coach somewhere in the meantime – this was a not unknown scenario. The exact identity is not known for certain, but it could be SR No. 446, noted in carriage registers as being grounded, but no location is given. This van was built by the LCDR at Battersea in 1897, as their No. 1061, later SECR van No. 560. It lost its side lookouts (where the plain steel sheeting may be seen) in 1923 and was withdrawn in October 1931. However, it could easily have been one of the numerous departmental conversions of this diagram (SR Diagram 879) that took place between 1926 and 1934, in which case the date of grounding would have been rather more recent. SR van numbers ran from 418 to 463, but as this example was oil-lit, it could not be in the range 450-463 as these had gas lighting. *A. E. West*

The equivalent ex-SER/SECR guard's vans could also be grounded. This is SR No. 467 (ex-SECR No. 106); an example of safe-fitted Diagram 881, built in 1909 for bullion traffic between Folkestone/Dover and London. The SECR had about half a dozen vans so equipped, although the safe was nothing more sophisticated than a box-like enclosure extending right across the vehicle at one end, accessible by the use of a standard carriage-door key. This was just like a lathe chuck key – and the writer, in his commuting days, always carried one in his brief case – for use accessing a guard's or luggage compartment when the existing passenger accommodation was full. It came in useful on overcrowded trains during the various rail strikes!! The safe is at the far end, but is largely obscured by the joinery leaning against the van. However, the running number may be seen and is in 12-inch white figures on a red panel – in order to make this conspicuous to railwaymen but not to the general public. This number may be seen on the second panel in from the right-hand end. The bullion traffic ceased during World War 2 and the vans were withdrawn in July 1940, after which this example was grounded at Eastleigh Works. The rest of the livery was standard olive green and traces of this may still be seen. When the traffic resumed, replacement utility vans in the form of SR Van C safe-fitted vehicles numbered 10-14 were provided, but had their number displayed in exactly the same manner. Once painted BR crimson lake, the backing panel was altered to blue, reverting to red once Southern Region green livery was reintroduced after 1956. *D. Cullum*

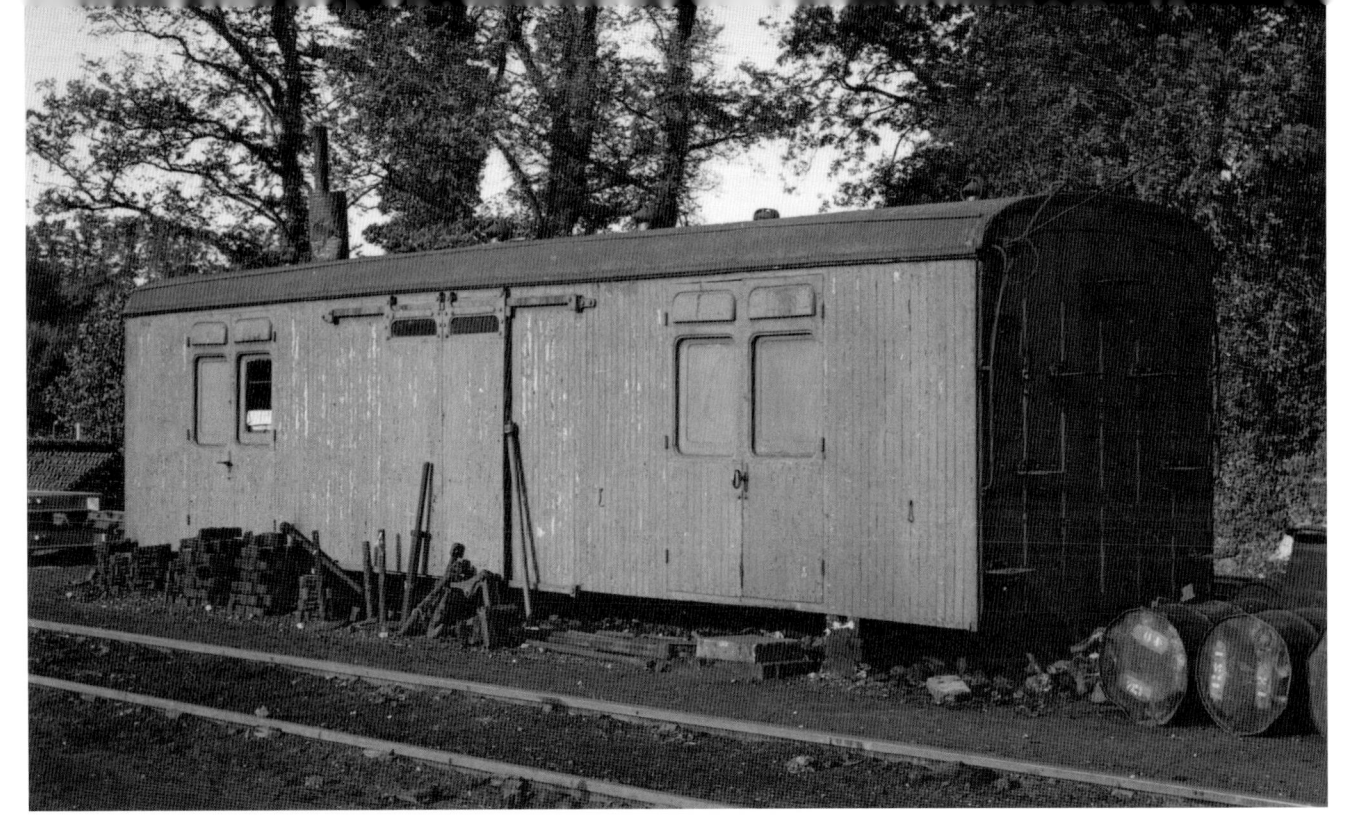

This completely matchboarded van to SR Diagram 883 was grounded at Shepherdswell in July 1941. It was built as SECR No. 613 by Ashbury Carriage & Iron Company in 1900, but was then fully panelled in the usual SECR style with lookouts at all four corners. It was probably rebuilt in the form shown sometime between 1917 and 1923, at a time when Ashford Works often used vertical matchboarding for partial or complete re-sheeting of a number of vehicles. This one appears to be a unique example of the diagram and was actually allocated SR Diagram 883A to distinguish it from the others. As SR No. 505 it was finally grounded as seen and remained on site until at least 1960. The picture was taken on 25 September 1948. Its dimensions were 32ft long and 8ft 4in wide over sliding doors. The stove chimney has had the addition of an old oil barrel but whether this was to aid the draw on the fire, remedy a roof leak or mischief is unknown! *J. H. Aston*

With shingle in the foreground, the sea behind and a Folkestone-registered fishing boat on the right, this can only be Dungeness beach. It is also another example of a partially matchboarded vehicle. This is ex-SER ventilated fruit van SR No. 1801 – one of two grounded at Dungeness in early 1933 (the other was No. 1815). There were 35 examples of the type built for Continental 'Grande Vitesse' perishable traffic from Folkestone and Dover to Ewer Street depot (near London Bridge station) between 1888 and 1897 and they remained on this traffic until withdrawn between 1930 and 1934 – SR Diagram 953 being allocated. They were 21ft long and 8ft wide over body. The picture dates from 30 August 1949, by which time the body had been tarred over, but the former number could still be seen under the final coat of paint. It was still there in the 1960s, but derelict. The winch alongside was used by the local fishermen to draw their boats up the beach and very often one could buy their catch direct from them alongside their boats. The writer can vouch for the locally-caught dabs (the only time he ever tasted such fish). *D. Cullum*

Down to Earth Part 4

The final picture was also taken at Dungeness on 30 August 1949 and features an ex-LCDR horsebox – a rather unusual type of vehicle to be grounded. But remarkably there were at least six grounded at this location in 1922-24 and the end of another may just be seen on the left, built into a shed. Perhaps they were used for stabling purposes – indeed, what better way to get around the Dungeness shingle than on horseback! By the 1970s, these two were just down to skeleton frames but in this 1949 picture the former SECR livery may be clearly seen. One other was grounded in Hythe, while others went to Ashford, Beckenham Junction, Margate Sands, Dover, Hastings and Herne Hill sidings. So, out of 35 examples, no less than 17 found further use! Not one became SR stock as all were condemned before Grouping but, in comparison, out of an inherited stock of 720 other Southern horseboxes not a single example has been recorded as being sold off or grounded. However, there might be an explanation. Of the 17 ex-LCDR examples grounded, no less than seven were purchased by a 'Mr Sinclair' - including those at Ashford, Hythe and five of the seven at Dungeness – and he was an SECR employee at Ashford Works who, maybe, had a sideline in horses? Perhaps their diminutive size (just 14ft 1in long) made them an attractive proposition. LCDR numbers were 1-35, later SECR Nos. 184-218, and they were built in two batches in 1879 and 1882-84. Denis Cullum fortunately sketched and measured up the grounded body at Herne Hill (on 7 June 1948), enabling the writer to prepare his own drawing – but like the SER composite mentioned earlier, it took until 1991 to make use of the sketch!

There is an interesting (and perhaps unfortunate) postscript about the vehicle grounded at Hythe (Kent) – horsebox LCDR No. 1/SECR No. 184. This went to the station as accommodation for the yard checker in 1922 and was sold a year later to the mysterious Mr Sinclair, but appears to have remained at Hythe station. It was resold in 1937 and moved to a house nearby – and the writer was told this by a friend of his great aunt (who retired to the town in 1952) and she lived a few doors down the same road (at 41 Tanners Hill Gardens). The vehicle was then in use as a garage at No. 1, with one end removed completely. The Bluebell Railway became aware of this body around 2007 and removal to Horsted Keynes was duly completed in November of that year. Regrettably, a bonfire disposing of scrap timber got out of control in 2010 and consumed this body (and an ex-LBSCR covered van body as well) – leaving the railway with just the nicely paint-stripped ironwork. As most of the timbers were by then rotten, they would have needed replacement anyway, so the spectre of rebuilding is still possible – although as a non-passenger-carrying vehicle this is not likely to be seen as a high priority. However, in view of the uniqueness of the vehicle, perhaps this ought to be reconsidered?

Readers may have noted that many of the pictures used in this series come from the collection of the late Denis Cullum. Denis was known as 'Coaches Cullum' amongst his circle of enthusiast friends (which included Dick Riley, Jim Aston, Ken Carr (Mr Pamlin Prints), Hugh Hughes and John L. Smith (Mr Lens of Sutton)) and he was also a railway employee. This writer owes much of his own knowledge of Southern carriage stock to Denis whilst his photographic collection includes many more SR coaches (as well as other interesting subjects) and is available from the present-day Lens of Sutton Association. *D. Cullum*

In contrast, just three 60ft 'birdcage' coaches were grounded – two brake coaches became Southampton staff social club buildings (possibly still on their wheels) in 1958 and the third was perhaps incidental. Many wooden-bodied coaches were broken up at Newhaven Cedar Sidings between 1944 and 1965 and the breaking-up gangs made use of Diagram 316 saloon composite coach 5465, which was grounded there in late 1958 – but probably only because it had been sent there for scrapping and the gangs noticed the comfortable first class saloon complete with armchairs so presumably decided to use it as their messroom instead! A distant picture of it appears in my 'Southern Rolling Stock in Colour' album (Crecy Publishing, 2015) – on page 25.

The one corridor coach grounded was a 'Continental' 62ft brake first – "presented to the children of Lambeth" (by whom?) in May 1959. This was Diagram 551 vehicle 7746 which had previously seen service in Folkestone and Dover boat trains. According to a press cutting (presumably taken from a South London newspaper and headed "The Lambeth Walk Express") it was grounded at the Lollard Adventure Playground Association's premises at Wake Street, Lambeth. The chair of that organisation, Lady Allen of Hurtwood, stated that it would be sited on the blitz-site playground and provide a boys' workshop, girls' meeting room and centre for old folks' tea parties and whist drives. How long it remained there is unknown, but the location still hosts a children's adventure playground today – but there is no sign of coach 7746. Unfortunately, the picture in the press cutting is not of sufficient quality to reproduce here, but it shows the coach on two road trailers negotiating a tight corner into Wake Street, still visibly painted BR Southern Region green with its BR markings and number clearly legible.

Many of the ex-LCDR six-wheeled and bogie coaches were sent to the Isle of Wight, where the final examples lasted until the late 1940s and some of these were grounded over there – enabling the Isle of Wight steam railway to recover some for eventual restoration. Pride of place must go to both coaches of pull-push set 484 which have now been restored and are in service at Haven Street. Several other ex-LCDR vehicles have also been restored and may be seen in service on the Island, on the Bluebell Railway and on the KESR. We will start by looking at some Chatham stock, before moving on to ex-SER four/six-wheelers and then finally on to SECR bogie coaches.

The next (final?) instalment will cover what few ex-Southern Railway post-1923 coaches that were grounded; electric stock vehicles – there were a few; the Isle of Wight; and other minor companies;, plus a few oddments that I (and my band of observers) came across on our travels.

Superheater Development on the Southern Railway

John Harvey

Drummond D15 4-4-0 No. 464, presumably in May 1915 having just been fitted with Urie's 'Eastleigh' superheater.

Introduction

The general run of railway literature and its authors, when discussing steam locomotive developments and performance in the twentieth century, might mention the benefits of superheaters or superheated steam, but discussion is generally absent. Sometimes the temperatures of superheated steam are tucked away in tables of data concerning locomotive performance, but commentators rarely explore the significance on economical working. However, it is clear that many Chief Mechanical Engineers (CME) and their close employees did understand the benefits; so this article has been prepared as an attempt to offer some explanations, hopefully without too much technical complication.

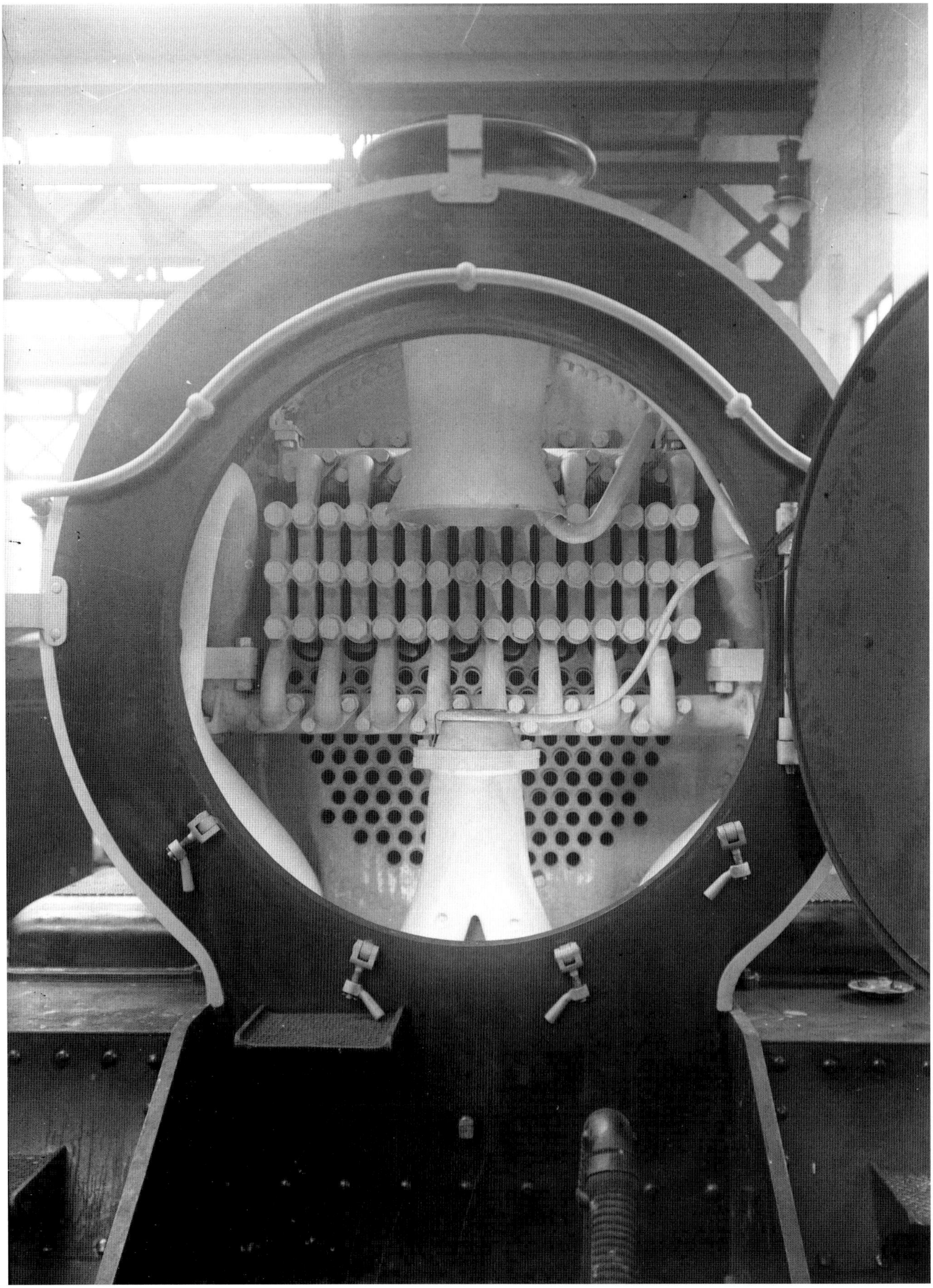

So, is a superheater just a superheater (?). For example, when the Ian Allan 'ABC' books state that locomotive class 'A' is superheated and that class 'B' is also superheated, is that all there is to it? After all, both classes are superheated so is there a difference between the superheaters? In design maybe, but as regards the steam temperature achieved it is easy, perhaps, to overlook the matter as being of little importance. The writer recalls a discussion he once had with a Great Western devotee who held G. J. Churchward in the highest regard but who did not appear to understand that his low temperature superheater, perpetuated by C. B. Collett, held inferiorities when compared with the designs of R. W. Urie and R. E. L. Maunsell on the SR. We will here try to explore the matter.

Measurements of the work done by a locomotive at the drawbar using a dynamometer car and knowledge of the quantity of coal burned and its calorific value enables the overall Thermal Efficiency to be calculated. With an express passenger locomotive this figure was usually no more than 10%. That means that less than 10% of the energy in the coal being burned is converted to useful work at the drawbar.

However, if we wish to assess how good the locomotive's valves and cylinders are at converting the energy in the steam to useful work, without worrying about heat losses due to the boiler or the losses due to friction in the locomotive and tender, the Rankine Efficiency can be calculated. The heat energy that is provided in the steam entering the cylinders less the heat energy wasted in the exhaust steam represents the theoretical maximum energy that can be converted to useful work by the engine. This calculation represents the core of the Rankine Efficiency calculation and it is evident that the higher the temperature of the steam reaching the cylinders, the greater the work that can be done for a particular mass of steam and the higher the efficiency. Actual efficiencies of steam locomotive cylinders were often compared with the Rankine figures to assess the quality of a design against the theoretical maximum (eg, 6, below).

Notice that the energy in the exhaust steam plays a significant part in the calculation and should not, therefore, be ignored when assessing overall locomotive performance. In a locomotive, the exhaust usually gets discharged to atmosphere (we will ignore exhaust steam injectors and the provision of condensers) and in the light of the well-known fact that at atmospheric pressure water boils at 100°C (212°F), the exhaust temperature of the steam is close to that figure. However, restrictions in the passages provided for the exhaust steam, ie, valves, pipework and maybe unduly high pressure in the blast pipe, cause an increase in back-pressure on the pistons, reduce power output and also cause an increase in exhaust steam temperature above 100°C (212°F). Poor design in these areas degrades power and economy. To minimise these adverse effects, good designs employ large valve openings for the exhaust, and multiple rather than single, blast pipes. However, such considerations are relatively small when compared with the benefits of superheating and so are outside of the scope of this article. The locomotive classes considered here are all 4-6-0s and each had a single blast-pipe and chimney: any differences in exhaust steam conditions between them will here be taken to be relatively insignificant.

Other losses such as the heat that disappears in the flue gas as it passes through the chimney to the atmosphere, and the frictional losses that occur, for example, in a locomotive's wheel bearings and valve motion degrade the Rankine numbers towards those of the overall thermal efficiency. Thus, in order to compare the Thermal Efficiency figure with the Rankine Efficiency, the latter has to be factored by the boiler efficiency, the actual efficiency of the cylinders and mechanical efficiency of the locomotive.

Information concerning superheater performance is not plentiful, but the following books have been used, particularly that by Harold Holcroft:

1) O. S. Nock, *The Southern King Arthur Family,* David & Charles, 1976;
2) O. S. Nock, *The GWR Stars, Castles and Kings,* Guild Publishing, 1986;
3) O. S. Nock, *History of the GWR Vol 3*, Ian Allan, 1967;
4) H. Holcroft, *Locomotive Adventure, Vol 2*, Ian Allan, 1965;
5) E. J. Nutty, *GWR 2-Cylinder Piston Valve Locomotives,* E. J. Nutty and Sons, 1977.

Further, an official report from:

6) CME Dept (WR Swindon), *Comparative Superheat Trials WR 'Castle' Class Engines* (October 1949) was used to obtain cylinder performance figures for GW locos of the same class.

The writer is grateful to Ben Carver for making available certain information relating to Robert Wallace Urie, CME of the London & South Western Railway.

Superheating Practice

Unlike his predecessor Dugald Drummond, Robert Urie recognised the economic benefits of properly superheated steam. First developed by the German engineer Wilhelm Schmidt in about 1890, the earliest application to steam locomotives was to the Prussian 'S4' class in 1898 which went into commercial production in 1902. George Jackson Churchward, CME of the GWR, demonstrated the benefits of the Schmidt superheater in 1906, but in the light of certain drawbacks and royalty payments for its use, Churchward developed his own design, culminating with the Swindon No.3 superheater in 1909. The cautious Churchward, who preferred to evaluate carefully every new development, was concerned that lubricating oils could break down under the high temperatures of superheated steam, causing increased wear and the requirement for increased maintenance of cylinders, valves and pistons. Churchward's superheater produced a relatively low degree of superheat, around 100°F (5) (less than 60°C), and this remained the GWR's standard until WW2. In 1913, Churchward claimed that 625 engines fitted with his superheater were saving 60,000 tons of coal per annum.

Opposite: **The Urie 'Eastleigh' superheater arrangement in the smokebox.**

Meanwhile Douglas Marsh, CME of the LB&SCR 1904-1911, tested two of his class I3 4-4-2T engines, one saturated and the other with a Schmidt superheater, resulting in the latter proving superior in both performance and efficiency. In comparative trials against current LNWR engines, the Marsh I3 class convincingly demonstrated the benefits of superheating.

If we step back and consider what a superheater is and what it does, in simplistic terms, as heat energy is added to boiling water at 100°C (212°F) at atmospheric pressure bubbles of steam will be produced. Both water and steam are at the same temperature. So, if the gaseous steam is drawn off from the water, as in a locomotive boiler albeit at an increased pressure, that gas will usually contain microscopic droplets of water which give the steam its white appearance. When that gas is free from the presence of water, but has not increased in temperature it is known as dry saturated steam and it is colourless. Since any water droplets present in steam drawn off from conventional locomotive boilers cannot do much in the way of useful work in the cylinders, engineers, for example Dugald Drummond, provided steam driers to change remaining water to steam. These usually consisted of additional tubes in the smokebox using left-over heat from the products of combustion after passing through the boiler as a source of heat.

A superheater adds energy to the steam, increasing its temperature and expanding its volume but there is, of course, no increase in pressure. It can be seen that, very approximately, if a particular volume of unsuperheated, ie dry saturated, steam produces a particular quantity of work in an engine's cylinders at normal working pressure, then the same volume of superheated steam will produce similar work. The point is that a smaller mass (weight) of superheated steam will be needed than saturated steam for the same work and this represents less fuel being used to boil the water. Hence the claim that Churchward made, above.

With Schmidt and others patenting their designs, the main attention became focused on methods of connecting the individual superheater tubes into the steam pipework between regulator and valve chest without infringing those patents and to avoid royalty payments. The basic approach was to provide two chambers called headers, one connected to the steam pipe from the regulator and the other to the pipes that fed the valve chests and cylinders. Superheater tubes were connected to the saturated steam header, passed into boiler fire tubes to be heated by the products of combustion and returned via 'U' bends along the same boiler tube to be connected into the superheater header. Unsurprisingly, railway companies went to great lengths to avoid royalty payments. Apart from Schmidt, who designed clamping systems to fasten the tubes, another well-known patent was held by John Robinson (who joined the Great Central Railway in 1900 and was CME 1902-1920) whereby the tubes were mechanically expanded into the headers. Urie, Maunsell and Gresley, to name but three, were each able to patent their own designs with the object of not infringing other patents. Such action was sometimes challenged by a patent holder, leading to contentious meetings and correspondence, and potentially to litigation.

By 1914 various types of superheater had become available. Schmidt produced a number of designs, one of which consisted of the saturated and superheated chambers arranged side-by-side and separated by a dividing wall in one casting to form the headers. The superheater tube ends were clamped in place. Richard Maunsell worked with his personal assistant George Hutchinson on a similar arrangement at Inchicore Works in Dublin, successfully filing their patent application in February 1913. Here, the tubes were fitted with ball-jointed ends (not unlike some plumbing fittings of today, where an 'olive' is slipped on to the end of a piece of pipe with a nut to tighten it into a socket) and a clamping plate covering a pair of tubes. A 'tee' shaped bolt was arranged with its head securing the clamping plate. With the bolt passing through a widened section of the dividing wall to a nut, the assembly could be pulled up 'tight' with the ball-joints snug in recesses in the header. This was the well-known 'Maunsell' superheater. Ease of maintenance on individual elements was provided by undoing the securing nut and releasing the 'tee' bolt and clamping plate after which the tube set (element) could be withdrawn from the boiler.

Neither Urie nor Maunsell appeared to be unduly worried by the breakdown of lubricating oil under high superheat temperatures. Higher temperature oils were being introduced. Urie gave trials to the Schmidt and Robinson superheaters on his own design of H15 4-6-0 and although the merits of the principle of superheating were clearly demonstrated, these designs were deemed not without maintenance difficulties, let alone the nuisance of royalty payments. Urie therefore designed his own 'Eastleigh' type of superheater, which was patented in May 1914 and went into production at the end of the year with the first application being to the rebuilt H15 4-6-0 No. 335.

This appears to show a Maunsell superheater, but detail of the header is obscured by the petticoat below the chimney.

Superheater Development on the Southern Railway

Urie's design was relatively complicated, heavy and expensive in manufacture, but gave very satisfactory service. In essence, the saturated and superheated headers are quite separate. Pairs of distributing chambers made as one casting formed connecting pieces between the headers. The chamber forming the right-hand of the pair was connected to the saturated header and that on the left to the superheated header. The superheater tubes are connected to these distribution chambers by patented hollow bolts with holes that enabled the steam to pass firstly from the saturated chamber into the tube. Two up-and-back passes inside the boiler fire tube were then made before the steam entered the superheated chamber via a similar hollow bolt. Each pair of distributing chambers served three sets of superheater tubes (or elements) and there were eight sets across the boiler, making 24 in total. Around 150°C (270°F) of superheat was obtained. In the event of an element becoming defective, then a pair of distribution chambers with three sets of elements had to be removed for access.

When the Southern Railway was formed and Richard Maunsell became CME, the Eastleigh superheaters were gradually replaced by Maunsell's own type. Apart possibly from the dislike of 'not invented here', Maunsell's design was cheaper to manufacture and easier to maintain if not, perhaps, quite as reliable in service, although detailed information on the latter is lacking to the writer.

Before the Southern Railway was formed, during 1922 tests were carried out on the Maunsell rebuilds of the 4-4-0 classes D and E, which had become D1 and E1. The average superheat temperature was found to be mostly in the range 650 – 670°F (343 – 354°C). There was no dynamometer car and indicator diagrams were not taken, but the coal consumption was measured and on the E1 was found to be in a range from slightly less than 0.10 pounds per ton-mile to slightly more than 0.11 pounds per gross ton-mile.

It might, perhaps, be noted here that both the Maunsell and Urie designs of superheater produced similar degrees of superheat and significantly more than Churchward's. The Swindon No. 3 had two rows of boiler fire tubes each (normally) with three sets of superheater tubes making only a single 'up-and-back' pass in each set. The Urie and Maunsell designs each contained three rows of tubes with two 'up-and-back' passes in each tube.

Actual Performance

Now, in 1924 the GWR's *Caldicot Castle* achieved coal consumption figures of 2.83 pounds per drawbar horsepower hour (lbs/dbhp hr) when on test. The equivalent figure in terms of indicated horsepower (ihp) was 2.1 pounds of coal per ihp.hour. At the time, this was considered a remarkably good figure, so good in fact that in some quarters it was regarded with disbelief. It is interesting, then, to read H. Holcroft's analysis (4) of tests with the SR's new *King Arthur* class loco No. E451 *Sir Lamorak* in 1925. The relevant points were repeated by Nock (1) that the SR 2-cylinder 4-6-0 achieved 2.2 lbs coal per ihp hr. The SR did not possess a dynamometer car and so drawbar horsepower could not be measured, but the loco was fitted for the taking of indicator diagrams. The train loading was broadly similar to that of *Caldicot Castle*.

No. E850 *Lord Nelson* at Nine Elms with indicator shelter. *John Harvey Collection*

A similar but this time front view of No. 850 so adorned. Forward vision for the crew on the footplate must have been seriously affected.
John Harvey Collection

The Southern engine was burning coal some 7% lower in calorific value than the GW, and it might be expected that the mechanical losses in a 2-cylinder engine would be more than a 4-cylinder version. Arguably, these factors counted against E451, so what other factor(s) enabled it to perform so efficiently?

Published figures (3) for the test of *Caldicot Castle* show that the superheater outlet temperature only rarely exceeded 520°F (271°C) (around 120°F superheat), and that was when the engine was working hard. This is in line with Ernie Nutty's view (5). A calculation of the Rankine Efficiency of an engine working with this degree of superheat barely reaches 20%. Now, with E451 where Holcroft (4) quotes the superheater temperature as an average of 560°F (293°C), the Rankine Efficiency is at least 22%. Further, E451 was able to sustain a very useful 1,100 ihp and coal consumption was 0.073 pounds per gross ton-mile. In 1927, the superheat temperature measured on E850, the first of the *Lord Nelson* class, was found to be in the range of 570 – 610°F (299 – 321°C), the Rankine Efficiency for such a temperature being around 25%.

No wonder, then, that the SR Engineers were pleased with the power and economy of E451 and E850. Gresley, on the LNER, also realised the necessity of good efficiency in providing the A4 locomotives that were capable of through running from King's Cross to Edinburgh on a single tender of coal and Stanier, a former Great Western man, applied similar ideas on the LMS. Ideas for improved performance of the GWR *King* class locomotives were prepared at Swindon by 1938, but those in authority at the time appear not to have approved and it was only in the 1940s that larger superheaters were designed and fitted to a few engines.

To return to *Caldicot Castle*, the measured thermal efficiency was 8.22% (3), the boiler efficiency was published as 79.8%, the mechanical efficiency of the loco 74% and whilst the cylinder efficiency against the theoretical Rankine is not quoted in (3), elsewhere *Castle* cylinder efficiency has been found to be around 70% (6). These figures give a Rankine Efficiency of 19.9%. Possibly overawed with the 2.83lbs/dbhp hr. figure, the GWR authorities seem not to have appreciated the massive cost savings that they might have achieved had a higher degree of superheat been adopted in the 1920s.

However, back in the 1923-25 period, a Robert Urie design had been adapted with the successful SECR 'N' class valve motion and draughting, to produce the 'King Arthur' class. It is difficult to avoid the conclusion that before World War 1 Urie and Maunsell were 30 years ahead of the GWR when it comes to superheating!

An earlier version of this article has appeared in the 'Southern Notebook', journal of the Southern Railways Group.

Buriton Chalk Pits

John Perkin

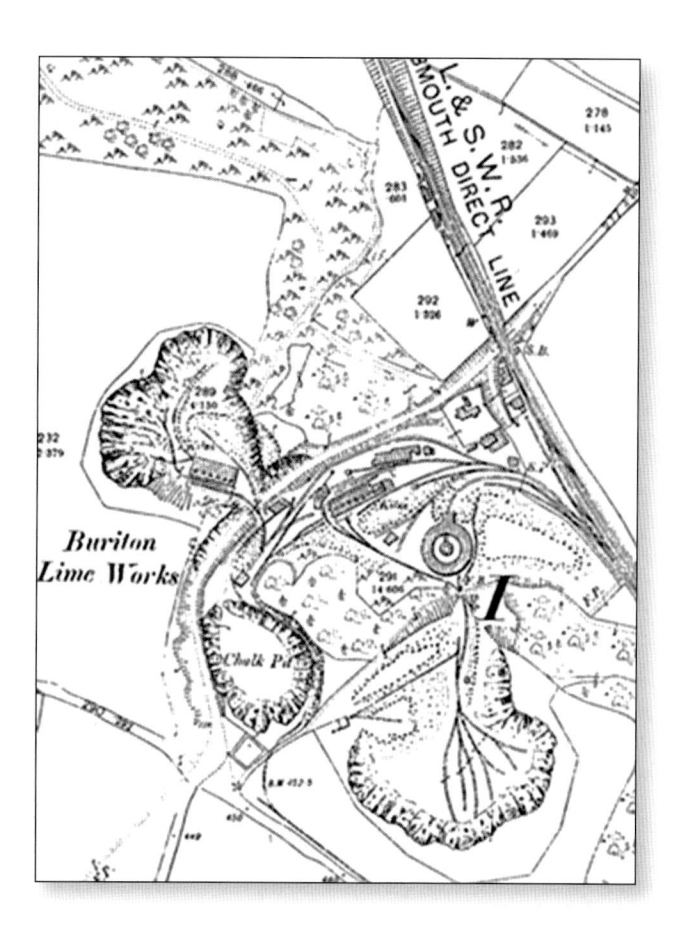

The earliest plan of chalk pits at Buriton in 1897. The Portsmouth Direct runs to the east of the workings.

The London to Portsmouth (Direct) railway line, opened in 1859, facilitated the introduction of the chalk pits and lime works at Buriton (four miles south of Petersfield). The works were opened around 1860 by Benjamin Forder. Later it became part of British Portland Cement Manufacturers Ltd. and was closed in 1939. (Mr Forder also had dealings with another lime quarry in Hampshire at Old Burghclere. Its history very much mirrors that at Buriton.)

History

For over 70 years until the Second World War, there was a significant source of activity in the chalk pits and lime works in Kiln Lane, Buriton. Chalk was excavated and burnt before despatching to factories for the manufacture of mortar and cement for the buildings industry. Lime was also used in agriculture.

There were two pits, one each side of the road known as Halls Hill, these were under lease from John Bonham Carter. The 1871 census records Mr Forder as a lime burner, employing twenty men and four boys. A lease from November 1878 also records that there was a recently constructed lime kiln. By 1899 there was a further lease for digging chalk, this between John Bonham Carter and Keeble Brothers of Peterborough. (The name Keeble Bros. is not referred to subsequently.)

At the time of the death of Lothian Bonham Carter in 1927, the works were leased to the British Portland Cement Company (BPCC) and provided, according to the sale catalogue for the Estate, "a substantial annual income." Perhaps it was this that partly persuaded the BPCC to purchase the site and they continued to operate the works until 1939. Apparently, the BPCC had hoped to convert the quarry to a cement works but this did not prove to be possible. Today much of the former excavations have been filled with waste.

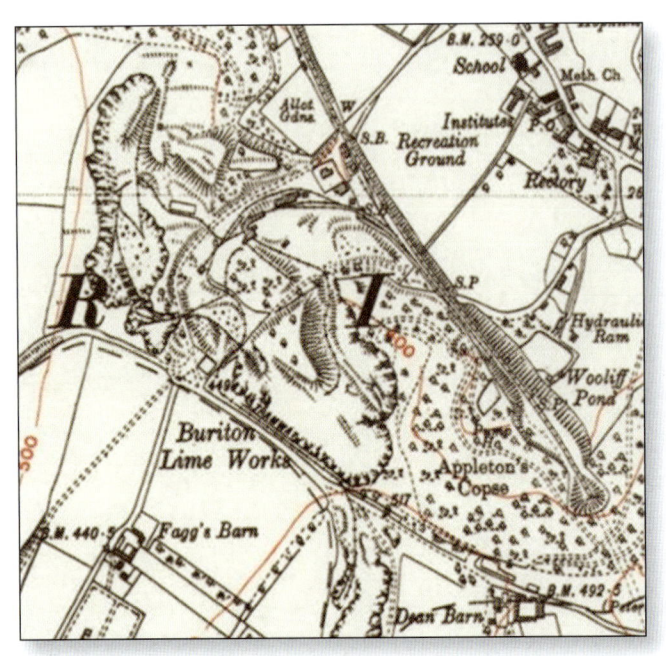

A slightly later plan from the early days of the twentieth century indicates development and expansion that has taken place in the previous years.

Similar date to the previous: note the position of the 'signal box' and interchange sidings.

The Three Chalk Pits

At Buriton three pits to quarry chalk were developed over the years, identified as 'France', 'Germany' and 'White'. 'France' would eventually became the local amenity tip in Kiln Lane. The 'Germany' pit was across the road south and east of Kiln Lane, and 'White' further south on Head Down. This latter site was later occupied by the Buriton Saw Mill.

A final plan, this time from 1932. The position of the trackwork within the workings should be regarded as 'a moment in time' as quarry lines would be moved as necessary dependent upon the site of the actual working although consistency was provided near the kilns and buildings.

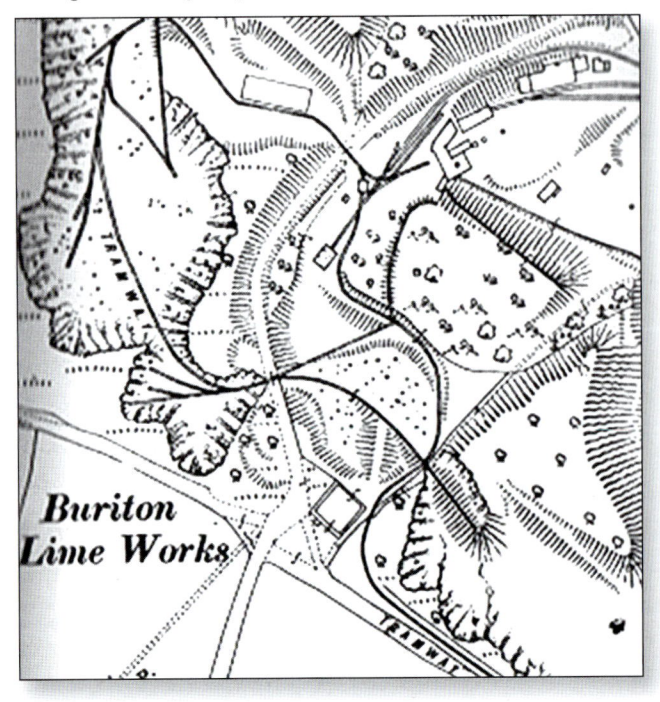

Maps of 1870 show a relatively small area of quarrying but by 1897 both the 'France' and 'Germany' pits are clearly visible with kilns in each. The 1932 map shows much more extensive workings including the extension into the 'White' pit.

The chalk in 'France' and 'Germany' had a higher clay impurity and produced a grey chalk which burnt to a creamy powder. This was used to produce mortar and, in later years, a very good waterproof cement.

The 'White' pit produced white chalk which was sold for the manufacture of plaster and was also used for gas and water purifying. The waste, also known as 'muckle', was ground and used for agriculture.

Lime Kilns

Different burning processes were evolved for use in the Buriton chalk pits. Flare kilns were used in the 'France' and 'Germany' chalk pits until around 1920. There was also a 'Hoffman' type kiln in 'Germany' which consisted of a circular group of kilns with a single tall chimney in the middle. This chimney was demolished in 1948. Later one main set of 'continuous draw' kilns were used. Some lime was ground in a mill initially driven by a stationary steam engine but later replaced by a diesel engine.

Range of crafts and skills

The lime works regularly employed about forty people, although this number more than doubled to nearer one hundred in the summer months when there was a greater demand for lime from the building industry. Men were employed around the clock, especially those necessarily tending the kilns, most of these individuals invariably local and during daylight at least would return to their homes for their mid-shift break.

BURITON SIDING (1938)

L & SWR NON-STANDARD BRICK BOX OF EARLY DESIGN.
OPENING DATE NOT RECORDED.
STEVENS' FRAME, 4½" CENTRES, 14 LEVERS.
PREECE 3-WIRE OPEN BLOCK.
CLOSING SWITCH:- ONE

Crossover Points 9 abolished pre-1960.

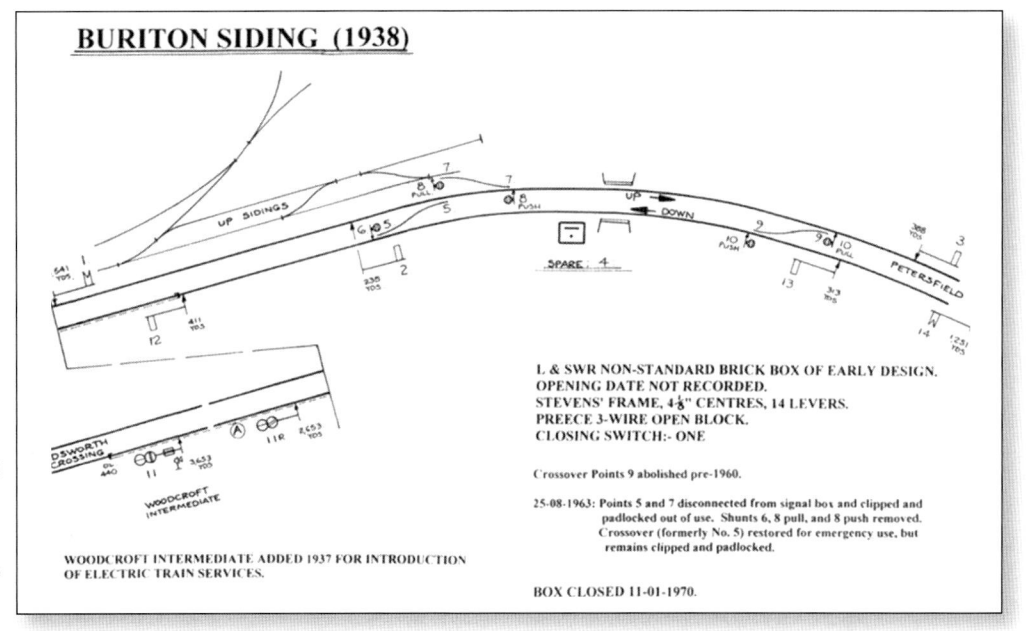

25-08-1963: Points 5 and 7 disconnected from signal box and clipped and padlocked out of use. Shunts 6, 8 pull, and 8 push removed. Crossover (formerly No. 5) restored for emergency use, but remains clipped and padlocked.

WOODCROFT INTERMEDIATE ADDED 1937 FOR INTRODUCTION OF ELECTRIC TRAIN SERVICES.

BOX CLOSED 11-01-1970.

Signalling plan c1938. Trains would call 'as required' and it will be noted the position of the crossover roads allowed for running round as necessary. It is not known if the facility had their own private owner wagons.

One person was employed to look after the horses used to draw wagons on a narrow gauge railway. There was also a granary on site to store oats for their feed. At the end of each day, the horses would be ridden, bare-back, down Kiln Lane and through the High Street by pony boys to be washed in the village pond.

Underneath the granary was the carpenter's workshop, the wooden 3ft gauge trucks for the internal railway were built and repaired here. Beyond the carpenter's workshop was a blacksmith's shop. There was also a bricklayer and a sail-maker, the latter to make tarpaulins to cover the lime in the railway trucks.

The lime works office was the northern part of one of two low sheds. Between them and the railway line was the old stables building and closer to the line was the sack mending shed. A little further towards the railway tunnel was the stores building with "Forders Lime Works" painted in white on the tarred galvanised iron roof facing the tracks. Two women were also employed on the messy task of sewing and mending hessian sacks.

The tall chimney referred to (also known as the shaft) was demolished on 14 August 1948 under the supervision of Jim Winser of Weston Farm and his business partner, a Mr Sykes from Froxfield. This pair had bought the lime works but decided that 'the shaft' was dangerous. All insurance estimates obtained by the new owners were rather high due to the need to cover the risk of anything affecting the nearby railway. With the decision made for the chimney to be removed, they formed a £100 company which would also have put an upper limit on any compensation available in case of an accident. Actual demolition was achieved in the same way as shown by the late Fred Dibnah in his various television series when demolishing chimneys from mills and suchlike. Some bricks at the base were replaced with railway sleepers which were then surrounded by wood and set alight. When the sleepers eventually burned, the chimney fell in the desired direction. Many of the bricks were subsequently salvaged, and cleaned ready for re-use.

A cruel version of the non-standard signal box likely dating back to the earliest days of the facility. Wagons of fuel for the kilns would arrive in the siding probably from a designated coal/coke merchant or colliery. Lime, however, was sent out probably in railway owned wagons in sacks and sheeted over.

The importance of the quarry relative to local employment is indicated here in this nineteenth century view. The man in the bowler hat may well be the site foreman.

Kilns, probably at 'France'. These were likely also similar to those at Old Burghclere.

The stack or chimney at 'Germany'.

More recent uses

During (or after?) the Second World War 'France' was used for steaming out and detonating land mines and is now filled with refuse. 'White' pit was later occupied by Buriton Sawmills. 'Germany' became a scrap yard for a time until a complete change of use saw it as a base for a bakery.

The remains of any kilns have now long since been buried but several buildings survive, including the mill, granary, carpenter's shop, offices and stables. The area of both the Chalk Pits and Lime Works are now a Nature Reserve. ('France' was an outstation for the mine warfare department of the Royal Navy's Torpedo School at Portsmouth, HMS *Vernon*, until at least the 'sixties', if not into the 'seventies'.)

Narrow Gauge Railway System

There was a 3ft gauge railway for the works which opened in 1912. A complex internal network of railways and inclines was developed to carry the chalk around within the site. Trucks would be horse-drawn to the edge of an incline, controlled by a pony boy, then they were run down to the kilns by gravity.

Those from the 'White' pit were run through the top of 'Germany' pit, across Kiln Lane into 'France' pit, under an incline, back across a second crossing over Kiln Lane and into the works where they would be tipped on to the loading floor of the kilns.

Trucks were loaded with about one and a half tons of chalk using gravel forks and these were then transported to the kilns in either 'France' or 'Germany' as there were never any kilns in the 'White' pit.

A man would stand on the back of the truck with his foot on the brake handle and another man would indicate when it was clear to cross Kiln Lane. The fuel was originally hauled up to the kilns by horses but in later years a stationary steam engine, cable and windlass were used to pull the railway wagons of coke breeze (undersize material not suitable for use in smelting furnaces, etc.) up to the loading floor level. Over the years, the cable cut grooves into the brickwork corners of some of the buildings as the wagons were hauled up the incline. There were also catch points to stop any runaway wagons.

Once at the loading floor level, the end door of the wagon would be opened, part of the load of coke would fall out and the rest would be shovelled out.

Locomotives

At least two Motor Rail Simplex machines with Dorman 40hp petrol engines are known of. These were to a standard design by this manufacturer, both also likely to have seen service in WW1 transporting supplies to the Allied front lines and also for towing ambulance wagons.

Whether locomotives were used as far as the workings is unlikely, dependent upon the number of horses in use may also have meant the services of a local farrier were also retained.

The two engines were:

No. 407 was ex-WDLR No. 2128 of 1917, transferred from Magheramorne Works in County Antrim, Northern Ireland, in May 1925 and rebuilt for 3ft from the original 2ft gauge.

No. 449 was ex-WDLR No. 2170 of 1917 rebuilt for 3ft from the original 2ft gauge. Both were subsequently sold or scrapped at an unknown date.

These are likely to be later wagons; earlier versions being simple wooden boxes strengthened with steel straps/banding at the corners.

Connection with the main line

Working instructions appeared in the Appendix to the Book of Rules and Regulations, that for 25 July 1921 stating:

Portsmouth Direct Line

Buriton Sidings. – These Sidings are situated on the up Line side between Rowlands Castle and Petersfield, the points being worked from Buriton Siding Signal Box. Down and up Goods Trains perform work at the Sidings under the supervision of the man in charge.

A loud sounding trembling bell, operated from Buriton Siding Signal Box, is provided in the Sidings, and five minutes before a Train or Engine is due to arrive or pass after being Block-Signalled in the ordinary way, the Signalman must switch on the bell and place or maintain the Shunting Signal in Forder's Lime Yard at "Danger." The bell must be allowed to continue ringing and the Shunting Signal must be maintained at Danger, until the Train or Engine has been brought to a stand at the Signal Box, or has passed into the section ahead. The gradient of the Main Line is 1 in 260 falling towards Petersfield.

(With grateful thanks to the South Western Circle, especially Bill Bishop, the Signalling Record Society and the Ordnance Survey. Further information may also be found at http://buriton.org.uk/history/buriton-lime-works/)

Belmont Hospital Bridge

George Hobbs

Towards Sutton with the final work proceeding towards the removal of the slab whilst below some sleepers still need to be added across the running line. *Images by the Author*

Epsom Downs Branch

In the early 1980s it was proposed to redevelop the site of the disused Belmont Hospital which lay to the west of the Epsom Downs branch between Sutton and Belmont stations. Originally the hospital had been the Metropolitan Schools for Orphans and when opened in Victorian days it had provided much traffic, passenger and freight, at Belmont station. Access to the hospital had generally been from the Belmont end but a second access at what is now Homeland Drive was via a single track lane over a narrow bridge, constructed with the branch in 1864. To provide adequate vehicular access to the new housing estate, replacement of this bridge was necessary. Meanwhile the Epsom Downs branch had been reduced to single track in October 1982, all traffic now using the former down line.

Removal of the old bridge took place over the Easter weekend in 1984. As there were no branch services on bank holidays or Sundays a four-day possession only involved suspension of services on Holy Saturday, and no disruption of commuter services. Except in extenuating circumstances construction noise is forbidden before 8:00am. At precisely that time on Good Friday, 9 April, the contractors detonated the charges to bring down the old bridge. The loud explosion was heard across south Sutton, giving a rude awakening to many local residents and acting as a 'starting pistol' for the four-day work. As is the usual practice with such work, a protective layer of old sleepers had been installed on top of the remaining track before demolition occurred.

During the possession, pedestrian access was provided by a temporary footbridge constructed on the south (Belmont) side of the bridge and this remained available until the new bridge was opened. Naturally the footbridge provided a good viewpoint of the work in progress. The first photograph, looking towards Sutton, shows the contractors working to remove the final parts of the old bridge. The lack of hard hats, ear defenders, high visibility jackets and safety boots would not be tolerated today. In the background a digger, engaged in unrelated work, is making use of the possession. The second photograph shows the old beams being removed by the contractor's mobile crane. Following clearance of the rubble from the old sleepers they were also hoisted clear of the line, as seen in the third photograph. Hard hats were deemed necessary by this time. The crane was situated on the east side of the line with easy access from Brighton Road.

Normal service on the branch resumed, as planned, on Easter Tuesday. As the bridge was to serve a new development the construction of the reinforced concrete replacement was more leisurely than if immediate restoration had been required. The bridge re-opened later in 1984 and now serves the housing development of Belmont Heights. The estate is easily accessed by foot at its southern end from Belmont Station, but frequent bus services on the Brighton Road mean that residents often prefer Sutton with its more frequent and wider range of services.

The replacement bridge was constructed to the full formation width so, although unlikely, reinstatement of double track would be possible. A plaque on the south parapet commemorates the bridge's replacement.

A road crane being used to remove the slab. In years gone by such work would have been undertaken by the railway's own cranes.

Opposite: **A collection of 24 sleepers being removed prior to restoring the line to traffic.**

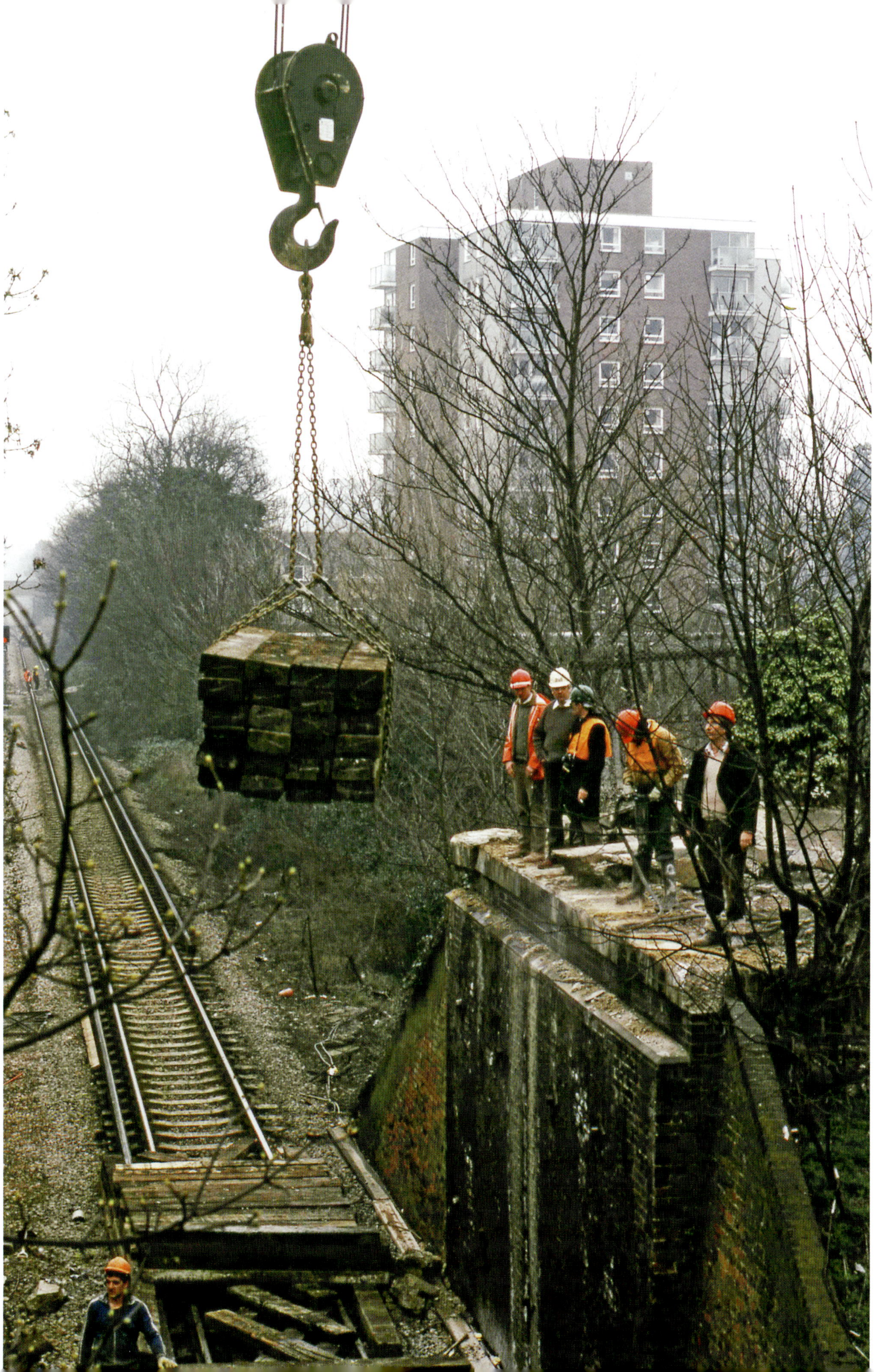

Branch Line Society Tour
7 March 1959
Images from the Stephenson Locomotive Society Collection
Courtesy of Gerry Nichols

From the feedback we have received, recollections of tours from days past have been a particularly popular topic in 'SW'. Continuing on the theme we are delighted to present a photographic feature on the Branch Line Society Portsmouth-area railtour of 7 March 1959.

The tour was noteworthy in that it encompassed four lines (or in two cases as much as remained) as well as a passenger working over the northern portion of the triangle at Portcreek.

Motive power was provided by M7 No. 30111 attached to pull-push set No. 6 (vehicles 6496 and 1103). Although we are not told the loading, just two coaches would have restricted the passenger numbers somewhat.

Starting from Portsmouth Harbour with No. 30111 leading, the special ran east to Havant and on as far as Fishbourne Crossing where the train reversed for the short run north as far as Lavant; the furthermost extent of the route to Midhurst. The return run was made this time all the way to Chichester, where the opportunity was taken to replenish the water supply.

No. 30111 and train at Lavant. The passenger service on this line had ceased in 1935 whilst its use as a through route to Midhurst was curtailed in consequence of a bridge wash out in 1951.

Branch Line Society Tour

Above: **Replenishing a thirsty engine at Chichester. The signalman has already cleared the starting signal for the run west to Botley.**

At Botley, probably on the outward run. The train would have arrived in the left hand platform and then shunted via the crossover to the position seen here. Five minutes were allowed for this move hence it was likely the view was taken at this time. The Bishops Waltham branch diverges to the right under the bridge. A stub remains as a headshunt for stone trains in the twenty-first century.

At Bishops Waltham terminus – another branch that was a casualty in the 1930s although freight, principally coal, continued to be handled until the early 1960s. Folklore has it that a number of local residents turned out to see what, to many, was the first passenger train at their station for many years.

This time the view is from what was then the A333 (now the B2177). This whole scene has now been completely obliterated with the station site replaced by a roundabout, although a level crossing gate and plaque exists at the start of a short nature walk on the right hand side towards Botley. (Unfortunately an illustration of a train on the nearby commemorative board shows an Isle of Wight O2!)

At Botley again, next stop Fareham. From this angle the miss-match between the two vehicles making up the PP set is clearly visible. No. 6496 is of LSWR origin whilst No. 1103 (nearest) dates back to SECR days.

Water in what was the former Meon Valley bay at Fareham. The circular indicator under the starting signal would display 'M' when cleared to indicate a Valley train.

Again participants and locals at Wickham. Passenger services on this route had ceased in 1955 at which time the centre section was also closed to all traffic.

The northern extent of the southern section of the Meon Valley line at Droxford. The station here, complete with canopy, has been beautifully restored in recent years.

Branch Line Society Tour

Fareham and the unusual sight of the starting signals 'off' for a train to Gosport. The repeating arms were a feature here due to the presence of the station footbridge. Upper quadrant arms had replaced lower quadrant signals a few years earlier.

The intermediate Gosport branch station at Fort Brockhurst; in the background is the level crossing over Military Road. Until the mid-1930s this had been the junction for the Lee-on-the-Solent branch.

Next came a long(ish) propelling move occupying some 50 minutes as the train ran west via Farlington Junction and Cosham Junction to Botley where there was a five minute wait. The participants may have wondered about the pause but all was explained in what was a comprehensive set of instructions from the Special Traffic Notice covering the working – this so far as the lines west of Portsmouth were concerned. (Similar instructions would have been published by the Central division to cover the passenger working to Lavant.)

Meanwhile at Botley, and having now complied with the working instructions, the train propelled to Bishops Waltham where a pause allowed photographs to be taken. It was now loco-first back to Fareham passing Knowle Junction on the way, also the divergence of the Meon Valley route but that could not be accessed at this point due to the track layout.

At Fareham there was a further reversal, the train now taking the tunnel route to Knowle and then pausing at Wickham before reaching its destination at Droxford.

Again after a suitable photographic pause it was back to Fareham (more water) and then south for the final branch to be visited, that to Gosport. The return was again via Fareham and thence back to Portsmouth Harbour, for the enthusiast a most rewarding way to spend six hours on a Saturday afternoon.

Gosport, which lost its passenger service in 1953, was a shadow of its former self following severe damage by bombing in WW2. The story goes that the station master, concerned about air attack, secured a new post at Wickham on the Meon Valley line and had his furniture and possessions loaded into a container ready to move the next day. Unfortunately that was the night the attack came and the station, along with his truck, were almost completely destroyed.

5

SATURDAY, 7th MARCH

BRANCH LINES SOCIETY'S SPECIAL TRAIN

No.	20	
	arr. p.m.	dep. p.m.
Portsmouth Harbour	A	12P33
	D	N
Havant	C 2 0	
Farlington Jct.		2 5
Cosham Jct.		2 7
Cosham		2 9
Fareham	2 19	2W22
Line	Single	
Knowle Jct.		2 27
Botley	2 32	2 37
Bishops Waltham	2 52	..
Bishops Waltham	..	3P 2
Botley	3 17	3 18
Knowle Jct.		3 23
Line	Single	
Fareham	3 28	..
Fareham	..	3W31
Line	Single	
Knowle Jct.		3 36
Wickham	3 42	3 50
Droxford	4 6	(4 15)

No.	21	
	arr. p.m.	dep. p.m.
Droxford	(4 6)	4P15
Wickham	4 31	4 32
Knowle Jct.		4 37
Line	Single	
Fareham	4 42	4W45
Fort Brockhurst	4 55	5 0
Gosport	5 20	..
Gosport	..	5 30
Fareham	5 58	..
Fareham	..	6P 6
Cosham		6 15
Portcreek Jct.		6 19
Fratton		6 24
Portsmouth & S'sea	6 27	..

A—Formed of engine and Pull and Push set. Pull and Push apparatus to be connected.
C—Via London Central District.
P—Propelling.

SPECIAL WORKING INSTRUCTIONS (12.33 p.m. Ex Portsmouth Harbour)

This special train will be worked throughout in accordance with the instructions under the heading 'Pull and Push Trains' shown on pages 56, 57 and 58 of the General Appendix to the Working Time Tables and pages 3 and 2 of Supplements Nos. 5 and 15 respectively, thereto.

BETWEEN BOTLEY AND BISHOPS WALTHAM

The instructions under the heading 'DURLEY LEVEL CROSSING' shown on page 150 of Supplementary Operating Instructions No. 7, will apply to the running of the special train between Botley and Bishops Waltham.

The special train must not exceed a speed of 20 m.p.h. at any point between Botley and Bishops Waltham.

All points over which the special train is required to pass in a facing direction and which are not provided with a facing point lock must be clipped and plugged in the correct position for the passage of the train and, where necessary, a Handsignalman appointed to control the movement.

The special train must run only to and from the platform road at Bishops Waltham.

Prior to the departure of the special train from Botley the Signalman must obtain by telephone an assurance from the person in charge at Bishops Waltham that the line to the platform road at Bishops Waltham is clear for the reception of the train. Before giving such assurance and prior to advising the Signalman at Botley of the departure of the special train on the return journey, the person in charge at Bishops Waltham must ensure that the route has been correctly set up and all facing points properly secured as required in paragraph three above.

On the return journey the Driver must be prepared to bring the special train to a stand at the 'from branch' shunt signal at Botley.

The bay road at Botley must be kept clear in connection with the running of the special train.

6
SATURDAY, 7th MARCH

BETWEEN KNOWLE JUNCTION AND DROXFORD

The instructions under the heading 'Meon Valley Line' shown on page 149 of Supplementary Operating Instructions No. 7 will apply to the running of the special train between Knowle Jct. and Droxford except as otherwise provided for below.

The special train must not exceed at speed of 20 m.p.h. at any point between Knowle Jct. and Droxford.

The special train must run only to and from the former Down Loop Platform Road at Droxford.

All points over which the special train is required to pass in a facing direction and which are not provided with a facing point lock must be clipped and plugged in the correct position for the passage of the train and a Handsignalman appointed to control the movement.

Prior to the departure of the special train from Knowle Jct. the Signalman must obtain, by telephone, an assurance from the persons in charge at Wickham and Droxford that the siding is clear for the passage of the train to the former Down Loop Platform Road at Droxford. Before giving such assurance the person in charge at Droxford must ensure that the route is correctly set to the former Down Loop Platform Road in accordance with the requirements of the preceding paragraph.

Prior to the departure of the special train from Droxford, the person in charge at Droxford must obtain, by telephone, from the person in charge at Wickham an assurance that the hand points leading to Mislingford Siding and Wickham Yard are correctly set and secured for the passage of the special train in accordance with the requirements referred to above.

The person in charge at Droxford must advise the person in charge at Wickham of the departure of the special train.

The Signalman at Knowle Jct. Box must advise the person in charge at Wickham when the Siding Starting Signal (No. 14) at Knowle Jct. has been lowered for the special train to proceed to the single line and that line is clear to Fareham East. The person in charge at Wickham must not authorise the Driver of the special train to pass beyond the Stop Board at Wickham until he has obtained this assurance from the Signalman at Knowle Jct., after which he must advise the Signalman of the departure of the special train.

FAREHAM—GOSPORT BRANCH

The instructions under the heading 'Gosport Branch' shown on page 150 of Supplementary Operating Instructions No. 7 will apply to the running of the special train over this branch.

Except where a lower speed restriction may be in force the maximum speed of this train must not exceed 20 m.p.h. at any point between Fareham and Gosport.

The train must run only to and from the Platform Road at Gosport.

Prior to the departure of the train from Fareham, the Signalman at Fareham West Box must obtain an assurance from the person in charge at Gosport that the line to the platform at Gosport is clear for the reception of the train.

Before giving such assurance and prior to advising the Signalman at Fareham West Box of the departure of the special train on the return journey, the person in charge at Gosport must ensure that all points over which the train is required to pass in a facing direction and which are not provided with a facing point lock, are clipped and plugged in the correct position for the passage of the train and a Handsignalman appointed to control the movement.

GENERAL NOTES

During the passage of the special train, shunting movements must be suspended on all sidings converging directly on to the line upon which this train will run.

Passengers may be permitted to alight, at their own risk, at the platforms at Fort Brockhurst, Gosport, Bishops Waltham, Wickham and Droxford.

(R.103312)

With grateful thanks also to the excellent 'Six Bells Junction' website.

Down to Earth
A Postscript
John Burgess

Back in the 'SW' office *(first floor overlooking the B4000 and the field of sheep)* we recently received a most appreciative email from John Burgess.

Rather than include it in 'Rebuilt' it is far better placed on its own, John writing as follows, "I'm enjoying reading the series of articles on grounded coach bodies, and wonder if the attached photographs might be of interest. In 1934, my maternal grandparents returned from several years working in Shanghai, and settled into life in South London. They regularly took holidays in West Sussex with their extended family, and in about 1934 or 1935 stayed in a property virtually on the beach at Pagham (which may still exist as a dwelling), which I believe consisted of two carriages laid out in a 'T' shape, of which I have found the rather charming view below. I think the older child is my mother, the end profile of one of the carriages is clear, albeit protected with timber weatherboarding. It's possible that a third vehicle might be hidden behind."

Speaking of the Meon Valley line as we were in the previous piece on the Branch Line Society tour, Roger Simmonds recently came across this view showing the railway at Privett probably just prior to opening in 1903. Although no train has run through here for nearly 70 years (the route has been closed longer than it was open), this is one of three out of the five stations on the line that survive as private dwellings. As in the best railway tradition, the location was also some little way from the village it purportedly served.

Three recent discoveries...

And speaking of correspondence, three recent discoveries from Roger Simmonds...

A stunning view of a family group on Shawford Down with the station of the same name (minus the 'Down') in the background. From the costume it probably dates from late Victorian times and so meaning the station is little more than a decade or so old – perhaps less – having opened in 1885.
R. Williams Collection

Finally a view of the vast number of railway staff based at Fullerton junction between Andover and Romsey during WW2. The numbers had been boosted by members of the SR Docks and Marine Office, evacuated to this rural backwater in consequence of enemy bombing. Here in peace and relative safety they continued their work based in various railway vans stabled on the stub of the Longparish branch. A special train was provided morning and evening to convey all between Southampton and Fullerton.

Christmas Mails and Parcels Traffic
A Retrospective Review Part 1
Richard Simmons

So where to start illustrating Christmas Mail and Parcel Traffic? At the time no doubt the last thing on the mind of any of the staff involved was to record their activities on film. Much of the working probably subject to frenetic bursts of activity associated with attempts to keep to schedules with both the GPO and the railway likely leaving the actual loading to the last possible moment. In turn this would have resulted in pressure upon the guard to get the train away so as to keep to its timetabled path, and in turn the locomotive crew to make up time where necessary and so keep to connections. Richard has kindly provided as wide a selection of images as possible for his piece but bear in mind almost all the actual workings described would not have been photographed; not least because at the time of year many of these trains would have run outside daylight hours. Here we see an exception, T9 No. 30707, on the Salisbury to Portsmouth & Southsea vans (via Southampton Terminus) recorded between Sholing and Netley.
All images by the author unless otherwise stated

Those of us of a certain age will doubtless recall the days when Christmas greetings were conveyed to recipients by means of Christmas Cards or letters, the majority of which were sent by way of General Post Office (GPO) postal services. Of some cards received maybe from the handwriting we were unable to recognise the sender's identity, and perhaps these circumstances presented us with something of a mystery which we liked to resolve before opening the envelope. Probably the clue to this conundrum could be found in the postmark which gave the vital clue as to the envelope's originating point. Remember, in those days there existed a myriad of postal sorting offices – some large, some small – dotted across the whole of the British Isles, each with their individual postmarks pointing to the area where letters were posted. And of course the majority of mail was transported by rail, sometimes by Travelling Post Office trains (TPOs) or by vans attached to scheduled passenger or van services.

Today means of conveying messages/greetings have vastly changed and more often than not exchanged by electronic means so diminishing the volume of cards posted. And alongside this reduction so too has the number of GPO – sorry, Royal Mail – sorting offices consequently reduced. But part of this reduction has been due to creation of large sorting offices not rail connected, seemingly serving very large areas of the country. For instance, when I moved to Farnborough in 1968 there was a sorting office there with its own individual postmark, but currently mail to and from this area is dealt with at a large office entitled "Jubilee Mail Centre" with no clue as to its location. (I have learnt since the actual location is at Feltham and can be seen from trains on the adjacent Windsor Line close to the site of the former Feltham Junction signal box, possibly even over, or partially over, the site of the long gone but not forgotten Feltham MPD.) Other large sorting offices I have come across includes one in southern Hampshire covering Southampton, Portsmouth and the Isle of Wight. In the latter case I suppose the area it covers is self-explanatory. Another Christmas Card I received bore the postmark proclaiming it was the mail centre for "BA, BS, GL, TA". Knowing the card's sender and their address I deduced this covered areas of Bath, Bristol, Gloucester and Taunton. I wonder how many sorting offices closed as a result of that creation! (We might even draw comparisons with signal box rationalisation, where individual boxes that once looked after 2-3 miles of track have been replaced by large area signalling centres.) Alongside this Royal Mail reorganisation, mail by rail conveyance has been vastly reduced from the levels conveyed in the last half century.

Having prepared an article for SW39, "Strawberries and Steam" covering Hampshire fruit traffic, readers may recall I ended that article with the comment on the basis that pursuing special traffic notices to gather details for that article had been most interesting but, "I would perhaps hold back from undertaking a similar exercise on the Christmas mails and parcels notice; now, they *were* beasts!" To that our editor added the caveat "for the moment...?" so here I am tackling it! Consequently this article is intended to examine how such traffic was handled on the former SR Southern area (see area detail below) between 1952 and 1969, those being the years for which I have relevant special traffic notices. Sorry if that disappoints readers resident in the former SR South Eastern and Central Divisions, but a large part of my railway career was spent on the South Western side of the region; consequently I have very limited access to Working Timetables (WTT) and Carriage Working Notices (CWN) of the other two divisions.

For administrative traffic purposes the South Western Division – which once extended from Waterloo to Padstow – was split into three divisions, London West, Southern and Western, with respective offices at Woking, Southampton Central and Exeter Central with officers in charge entitled Divisional Superintendents. In 1952 each was retitled District Traffic Superintendent; in 1961 District Traffic Manager; and then in 1962 District Manager; with the London West and Southern Districts being merged in 1963 into the new Line Manager, South Western Division set-up at Wimbledon. Even this organisation did not escape from being retitled further, becoming Divisional Manager by 1965. By Christmas 1963, of course, following regional boundary revision the Exeter district had become submerged into the WR Plymouth Division. In addition the London West and Southern districts lost their individual identity and became fully integrated into the aforementioned Line Manager South Western Division. Talk about going round in circles!

Each division produced its own printed Special Traffic (ST) notices, usually on a weekly basis supplemented with daily stencil notices containing revisions, etc, made after publication of the ST notices. Printed notices were consecutively numbered starting at 1 at the beginning of January with the prefix P and a suffix denoting division of production, with LWD for London West, SD for Southern and WD or Western District followed by year of production. The Somerset and Dorset produced their own notices from their Bath Green Park office.

Not surprisingly for an operation of this magnitude, notices prepared by three individual districts would have produced mayhem with all the innumerable loose ends needing to be tied-up. So one notice was prepared by the LWD – and numbered in their sequence – until 1962. From 1963, following creation of the Line Manager's organisation, the Line Manager took responsibility for production of the P notice which was numbered in their own sequence. I thought some explanation of the notice's background was both necessary and helpful. Having hopefully explained the procedure, now let's get on with the 'nitty gritty'.

I have managed to retain notices for 1952, 1953, 1956 to 1966 inclusive and 1969. After that year I ceased working in the operating departments which means I have no idea when production ceased, but I imagine it was somewhere around the time when Royal Mail's use of rail transport diminished and eventually ceased. Special arrangements for this traffic did not commence on the same date each year, generally starting around 13/14 December but in 1969 started as early as 8 December and terminating around 25 or 26, but in 1952 lingered until 27 December. In those years it will be realised special arrangements continued well after the last recommended dates for posting Christmas mail!

One aspect which could be relied upon every year, and in marked contrast to the present day situation, was that no major engineering work was undertaken throughout pre-Christmas periods. Also back in those days at least one collection was made from pillar boxes on Sundays. Special traffic P notices for this period contained between 57 and 87 pages per issue in the years under review. In addition it was customary for a proof issue of the notice to be produced and circulated to all concerned from which corrections and amendments were made. The final edition had a print date sometime between the last two weeks of November and first week of December. Underpinning production of the notice would have started relatively early each year with a series of internal departmental meetings in both BR and GPO organisations plus joint meetings between both parties, to determine special traffic arrangements and any specific

Christmas Mails and Parcels Traffic

staffing arrangements. Incidentally loading and unloading at stations of letter mail was undertaken by GPO personnel but GPO parcels was by railway personnel. In addition to special mail trains, the P notice included details of additional vans attached to regular van trains timetabled in WTTs together with additional vans attached to regular passenger services. Not all special trains ran for the full duration of the P notice, and others ran on Sundays only. Before reviewing individual P notices we might detail some notes on the type of vans used and Travelling Post Office (TPO) trains involved.

With the exception of a few parcels trains in the inner London suburban area which were EMUs, until the early 1960s all van trains were steam-hauled but very few pathways for associated light engines were shown. That would have been done by means of stencil notices (SNs) after engine diagrams and pathways had been determined. Whilst P notices included formation details of special trains, space constraints in this article prevent inclusion of details of individual train formations. So it will have to suffice to summarise types of vehicles used which include: BG (Brake Gangwayed), Cor PMV, (Corridor Parcels & Miscellaneous Van) PMV, PMV(4), Van B (General utility Van – differing types), Van B (Stove). Vanfits (Fitted goods wagon) were widely used. In the far west country, vans to some WR destinations, usually transferred to WR services at Exeter St Davids, were shown as WR vans or WR Syphon. Some services did not have the specific type of van quoted, merely being shown as "van(s)". There is no room to give detailed consideration to vans by ordinary services but the more interesting items will be highlighted.

So far no reference has been made to TPO trains of which the SR operated two, one from London Bridge to Dover Priory (South Eastern Train) and the other from Waterloo to Dorchester South, latterly to Weymouth (South Western Train). In view of their nature and history a more detailed consideration of them seems justified.

Down parcels train stopped at Winchester. Line occupancy requirements dictated that station work was dealt with promptly. When both letters and parcels were handled, letter sacks were dealt with by GPO staff and parcels by railway personnel. *S/Way collection*

As well as TPO vehicles, both trains included passenger accommodation albeit third class only in the case of the South Eastern train which was re-designated second class from the June 1956 summer timetable and subsequently standard class. For these services the SR was allocated nine Sorting (POS) and six Tender (POT) vans of which three Sorting coaches included a toilet. Three Tender (Post Office) coaches on the Dover coaches also included Guard's brakes. This information was obtained from the SR produced 'Appendix to Carriage Working Notices' which were published annually to commence with summer timetables, of which I have a copy for the years 1969 to 1973 inclusive. Allocation of these specialised coaches was as follows: two sorting plus one relief and two tenders and one relief to South Eastern trains, four sorting plus two relief and two tender and one relief to South Western trains. A caveat in the 1970 notice advised that two sorting and two tender vehicles each with toilets in South Eastern services would be replaced by vehicles "when heating questions were resolved". Replacement vehicle numbers were not quoted with an "S" prefix having an "M" prefix. The caveat was not included in the 1971 publication so presumably the aforementioned "question" had been resolved. In 1973 or early 1974, however, changes were made to the vintage of TPO vehicles. POSs 4920/1/2 (one traffic plus two spare/maintenance) were allocated to South Eastern services along with POTs 4958-4960, all with off centre gangways. A further revision took place during the 1973/74 timetable when the long standing use of London Bridge station for this train was transferred to Victoria, possibly because of what transpired to be commencement of the long term reconstruction period of London Bridge station.

Although this resumé is primarily (but not exclusively) involved with the South Western lines, we should not forget a similar amount of extra traffic was generated on the other divisions of the Southern - and of course on the other regions. This in turn resulted in demands for rolling stock and motive power far beyond the normal weekday requirements. Throughout the year there were other notable traffic peaks as well, notably Summer Saturday and Bank holiday traffic. The net result was rolling stock and locomotive power idle at other times. It was these peaks in traffic that Dr Beeching was later determined to remove and which in turn accelerated the move away from rail to road. (Where again there was more demand at peak times – and so on...) On the South Eastern section we see E1 No. 31507 on a down van train at Gravesend Central.

Christmas Mails and Parcels Traffic

I have public timetables from 1947 when the South Eastern train included passenger accommodation, when the down train departed London Bridge for Deal, Mondays-Saturdays at 11.50pm following the original South Eastern Railway route to Tonbridge via Redhill. Calling stations were East Croydon, Redhill, Tonbridge, Ashford (Kent), Shorncliffe – as Folkestone West was known by until the winter 1963 timetable but not at Folkestone Central – Dover Priory and Deal, and terminating at Dover Priory on Saturday nights. What the public timetable showed as a connecting service, again third/second class only, started from Ashford (Kent) at 2.30am for Margate calling only at Canterbury West and Ramsgate. It must be reasonable to assume that vans for this train were detached from the London Bridge train at Ashford (Kent).

Sunday nights saw an earlier start from London Bridge at 9.5pm, this time conveying first class accommodation as well as third/second. This was a slower journey than its weekday companion due to the number of intermediate stations at which it called: East Croydon, Redhill thence all stations to Tonbridge, Ashford (Kent), Shorncliffe, Folkestone Junction, Dover Priory, Martin Mill, Walmer for Kingsdown, terminating at Deal at 12.57am. As on weekday nights there was a connecting train from Ashford (Kent) to Margate starting at the witching hour and conveying both first and third/second class accommodation and calling intermediately at Canterbury West and Ramsgate.

In the opposite direction the train – again with only third/second class accommodation and running via Redhill – started from Margate at 9.28pm, called at all stations to Dover Priory except Margate East and Richborough Castle Halt (two stations long since closed), then Folkestone Junction, Folkestone Central but not Shorncliffe in this direction, Ashford (Kent), Tonbridge, Redhill, East Croydon arriving London Bridge at 1.19am, running on weekdays and Sundays. There was a connecting train both weekdays and Sundays from Margate to Ashford (Kent) via Canterbury West conveying first and third/second class accommodation.

Additional shunting engines – or an addition to the scheduled times an engine was available to shunt – was a feature of the Christmas traffic. Inevitably this also resulted in much more light engine workings than would otherwise have been the case. Waterloo had several pilot engines specially for this purpose, not just at Christmas but throughout the year as well. Here No. 82005 takes the opportunity for a top up during a quiet spell. (Question: was there not a balance pipe between the two side tanks? If so, why is filling taking place on the far side...?) *S/Way Collection*

From the public timetable it is difficult to deduce how far the Post Office vehicles operated, but from intermediate timings plus the fact that the down train terminated at Dover Priory on Saturday nights/Sunday mornings, I conclude these vehicles were detached/attached at that station during the week. Another interesting query; did these trains also convey vans containing Continental mails to be worked to and from Dover Marine from Dover Priory? Timings outlined varied little over the years and applied until the end of the winter 1962/63 timetable, when passenger facilities on both trains were withdrawn, so they survived electrification for at least a short time. But I am going to suggest that at least a contributory cause was withdrawal of Mark 1 loco-hauled stock. For instance, the May 1969 Appendix to Carriage Working Notices listed, apart from four 3-sets employed on the Tonbridge-Reading axis, only 18 carriages which were allocated to the SED as follows: 6 for "Golden Arrow", 7 for Night Ferry and 5 described as "diagrammed". Presumably "diagrammed" vehicles were employed on the two newspaper trains operating on the division, 3.0am Holborn Viaduct-Dover Priory via Faversham and 3.40am London Bridge-Hastings which shed a Dover Priory portion at Tonbridge.

Turning to the South Western, there were down and up TPO trains between Waterloo and Dorchester (now Dorchester South), whose origin can be traced back to at least 1914. The L&SWR summer timetable for that year (Ian Allan facsimile reprint 1967) included a train leaving Waterloo at 9.50pm for Dorchester, the heading at the head of the train's timetable column boldly proclaiming "Mail". In today's climate of security and terrorist prevention such a heading would not be permitted. Weekday calling stations were: – Basingstoke, Micheldever (June only), Winchester, Eastleigh, Southampton Town (reverse), later to become Southampton Terminus and which closed its doors on 5 September 1966, Totton, Brockenhurst and then via the original 'Castleman's Corkscrew' line to enable calls to be made at Ringwood and Wimborne, then Wareham and finally Dorchester arriving at 3.33am. Bournemouth was not totally left out as a portion was detached at Brockenhurst for Bournemouth Central, now known as just plain Bournemouth, with an intermediate stop at Christchurch. It is notable that Poole was not served. The Sunday train was a much more pedestrian affair not heralded as "Mail" and departing Waterloo at 9.15pm. It made little headway before making a "take up and not set down" (sic) stop at Vauxhall then calling at Surbiton, Weybridge and Woking. Beyond Woking it was all stations to Eastleigh except Brookwood (Necropolis) now Brookwood, Hook and Shawford. Then curiously at Northam followed by Southampton Town, Totton, Brockenhurst, where the Bournemouth Central portion was detached as on weekday nights, Ringwood, Wimborne, Hamworthy Junction – now just Hamworthy, Wareham and Dorchester before continuing on to Weymouth arriving at 2.53am at the end of a most tedious journey. This practice of continuing on to Weymouth with the Sunday night train continued thereafter.

On weekdays and Sundays the up train was not designated in the timetable as "Mails" and started its journey from Weymouth as it did in the BR era – and not Dorchester – at 9.50pm. Stopping stations were:- Dorchester then all stations to Wimborne except Broadstone Junction (in later years to become just Broadstone: and remember, Holton Heath was not opened until Southern Railway days in 1924). A Thursdays only stop was made at West Moors followed by Ringwood and Brockenhurst where a portion from Bournemouth Central was attached, Totton, Southampton West later to become Southampton Central, Southampton Town (reverse), Eastleigh, Winchester, Micheldever, Basingstoke, Winchfield, Farnborough, Woking, Weybridge, Surbiton and Waterloo arrive 3.35am.

It is interesting to compare timings and stopping patterns of these two trains with their modern counterparts.

At the end of steam in July 1967 the down train departed Waterloo weekdays and Sundays at 22.30 (by then, the 24 hour clock was in operation) calling at Woking, Basingstoke, Winchester (as Winchester City had become), Eastleigh, Southampton Central, Brockenhurst, Christchurch, Bournemouth Central, Poole, Wareham, Wool (early Monday mornings only) and Dorchester South except late Sunday night/early Monday morning when it ran through to Weymouth.

Hump shunting at Feltham – close to where the modern day Royal Mail Jubilee sorting office is located.

Christmas Mails and Parcels Traffic

The up train started from Weymouth nightly at 22.13 calling at Dorchester South, Wool, Wareham, Poole, Bournemouth Central, Christchurch, New Milton, Sway – on Saturdays only to set down, Brockenhurst, Southampton Central where it crossed with the 22.30 ex-Waterloo; TPO staff off this train changed with staff off the up train; then on to Eastleigh, Winchester City, Basingstoke, Farnborough – Monday mornings only, Woking, Weybridge, Surbiton, Vauxhall – to set down invoices for Nine Elms and Waterloo. In contrast to the South Eastern TPO neighbour, both South Western TPOs included 1st class accommodation.

When electrification commenced the down train departed Waterloo at 22.52 and the up train left Weymouth at 22.35, both diesel hauled. Both TPOs were withdrawn from the end of the timetable which was valid until 15 May 1988 although it had ceased running on Sunday nights some time before that, most likely at the time Royal Mail abandoned Sunday collections.

TPO Train Formations

Having described the TPO vehicles, it may come as a surprise to find that certainly in 1952 no sorting vehicles were included in either down or up TPO services in the Christmas period. Thus from Waterloo the down train was formed as follows:- 2 Corr PMVs for Bournemouth Central, 2 stowage vans gangwayed together, 1 BY, 3 Corr set (770 Bulleid type) for Dorchester South, 2 corridor thirds and 1 PMV(4) for Southampton Terminus, 1 Van B, 1 PMV (4), 1 PMV (4) with Lyons traffic for Portsmouth & Southsea.

The up train was formed:- 3 Corr set (770) and 1 Corr PMV Weymouth to Waterloo, attached at Dorchester South for Waterloo – 2 stowage vans gangwayed together, attached at Bournemouth Central for Waterloo, 1 Corr PMV, attached next to the engine at Southampton Terminus for Waterloo, 1 PMV (4) attached at Eastleigh from Portsmouth & Southsea for Waterloo, 1 van B and 2 Corr PMV from Plymouth Friary for Waterloo and 2 PMV (4) (see paragraph regarding connecting services for further details) also at Eastleigh from Dorchester South to Nine Elms and finally attached at Woking for Waterloo 1 PMV (4). An explanation is required as to how come the two PMV (4)s from Dorchester South to Nine Elms were attached at Eastleigh whereas the train had already called at Dorchester South. The fact is that they experienced a rather complicated journey from Dorchester South, starting from there on the 10.4pm by additional van train to Eastleigh, transferring there to the up TPO but only as far as Basingstoke, being transferred there to the 8.25pm Exeter Central-Walton-on-Thames additional van train and arriving at the latter location at 9.25am the following morning and finally finishing on the 2.35pm additional van train Walton-on-Thames-Nine Elms "J" shed arriving 3.10pm approximately 17 hours after leaving Dorchester. Small wonder the Post Office mantra was post early for Christmas!

Before finishing with TPO trains, whilst it may seem strange no actual sorting vehicles were conveyed in the immediate pre-Christmas period, it is very likely this was at GPO request because the sheer volume of mail conveyed at that time could have precluded complete sorting en-route. There seems little doubt that GPO sorting staff would have done alternative work and a clue to this comes from the fact that the P notice instructed that the two stowage vans on down and up trains were to be gangwayed together, thus indicating movement of mail bags between these coaches. Work done, maybe, by redeployed Post Office staff as the P notice also pointed out that Post Office staff supervised loading and unloading of mails.

TPO No. S49505 at Eardley sidings on 18 April 1960. The offset end gangway will be noted.

Van services connecting with TPOs

Some of these services together with mails also conveyed passengers as well as mails so a start will be made with the 1.23am MX Southampton Terminus/1.46am MO Eastleigh – Portsmouth & Southsea which was advertised in the public timetable only between Eastleigh and Portsmouth & Southsea. This was because in the days of mechanical signalling at Eastleigh, no running signals existed there for trains from the up Bournemouth line to run direct to the two down platforms, which had to be done by means of shunt signals, hence these could not convey passengers. This train conveyed vans for Portsmouth off the down TPO which included Isle of Wight traffic which was normally conveyed by road vehicles between Portsmouth & Southsea and Portsmouth Harbour stations for loading on to the 2.50am boat sailing across Spithead to Ryde Pier Head from where there was a 4.5am to Ventnor and possibly an unadvertised train to Cowes. Yet during the Christmas Mails and Parcels period, the train from Southampton Terminus/Eastleigh was extended to Portsmouth Harbour and back to Portsmouth & Southsea and in so doing succeeded in calling at Portsmouth & Southsea twice! The first occasion was at the High Level whilst en route to the Harbour and then running to the Low Level platforms upon termination.

In the opposite direction the 11.16pm (11.25pm Sundays) Portsmouth & Southsea-Eastleigh conveyed passenger accommodation and vans (some additional to booked working) with traffic for Bristol, Yeovil, Waterloo, Bournemouth Central and transfer for stations beyond these destinations.

Connecting at Eastleigh with these mail trains was a 1.55am Eastleigh-Yeovil Town passenger and mail train which also ran Sunday mornings, but not on Mondays. At Salisbury connection was made into a 2.55am MX Salisbury-Bristol Temple Meads which in the 1952 Christmas period was restricted to carrying passengers and letter mails. Earlier in the evening, however, there was what I term a timetable curiosity. That is, a train which did not follow the pattern of any other particular service nor did it have a comparable service in the opposite direction. This was the 4.40pm Plymouth Friary-Eastleigh which started from Plymouth North Road after Friary closed on 15 September 1958. The particular train in question most certainly did not fall into the express category, not arriving at Eastleigh until 11.48pm! This pedestrian feat was achieved by calling at all stations between Plymouth Friary or North Road and Salisbury except Tamerton Foliot, Bere Ferrers, Sutton Bingham, Dinton and Wilton South, but from Salisbury dashed non-stop to Eastleigh but even that was interrupted on Wednesdays and Saturdays by a stop at Romsey. It also conveyed two vans for Waterloo via Eastleigh. Connecting into this train at Salisbury was an 8.45pm from Bristol Temple Meads. The train from Plymouth was lucky to survive the cull of through trains from west of Exeter St Davids towards Salisbury in the post-1963 cuts, and its remnant can be found in the form of the 19.27 from Exeter St Davids in the reproduction of the September 1964 to June 1965 public timetable, part of Jeffery Grayer's interesting article 'Salisbury to Exeter Part 2' on page 35 of SW46. Although it may be viewed as a departure from the subject in question of Christmas mails traffic, I think it is interesting to record that in the West of England line's heyday, both the 1.55am MX Eastleigh-Yeovil Town and 4.40pm Plymouth Friary-Eastleigh were shown in the public timetable as recognised connections from and to Waterloo. In the down pages of the West of England table 50 for the 2.48am Salisbury-Yeovil Town (1.55am ex-Eastleigh), connecting times are shown for Waterloo, Woking and Basingstoke for the down TPO (10.30pm ex-Waterloo) plus even a connecting time from Surbiton into the train at Woking, with an explanatory column note "via Eastleigh", but makes no reference to a change of train being required at Eastleigh nor the long waiting time there from 12.53am until 1.55am. Similarly in the 4.40pm Plymouth Friary-Eastleigh column, arrival times are shown at Basingstoke, Woking and Surbiton together with the column note "via Eastleigh". Again, no reference to the long waiting time at Eastleigh from 11.48pm to 1.29am. It is difficult to imagine that at that time of night Eastleigh was a welcoming station for intrepid passengers, who must have been very few, making the change. Or was it to pacify any over-zealous nocturnal travelling ticket collectors that this was a recognised route for passengers from or to London and whom should not be charged an excess fare for incurring extra mileage to the direct route via Andover? But having undertaken a light hearted digression, let's return to Christmas Mails and Parcels.

Individual P Notices

The same train plan did not operate year in, year out and obviously economic and other circumstances necessitated alterations as years progressed. So I think the best way of conducting this review is to take the first year's notice (1952) as a template and compare with that, as well as possible, bearing in mind space constraints, the following years' notices. AM, PM and 24 hour clock times will be used in accordance with that in use in the year under review. Some station names changed with the passing of the years and names quoted will be those used in the year under review.

'Focal' Points

Several locations became, as it were, focal points, dealing with the many extra van trains and individual vans which could not have been accommodated at usual stations or sidings. One such location was Walton-on-Thames which many printed notices insisted on referring to as just "Walton", so in accordance with such notices, that name will be used. Sidings at Walton were, in fact, Oatlands Sidings located adjacent to the up local line (as slow lines were titled in those days), with entry controlled by Oatlands box and exited at the Walton end. These sidings were principally used to berth passenger stock – usually of older vintage – with such stock generally finding employment on summer Saturdays, Bank Holiday and other busy periods. Walton-on-Thames and Oatlands boxes closed on 22 March 1970 and the site of the sidings can just about be discerned where the cutting widens, but the whole area has been largely taken over by nature.

Slightly away from the text, but a South Eastern section image of No. 31402 on an up van train at Chislehurst. *S/Way Collection*

Another focal point was Teddington on the Kingston Loop where sidings were located adjacent to the down Kingston line between Hampton Wick and Teddington (whose signal box closed on 13 December 1966). Christmas workings were generally empty vans from Nine Elms 'J' Shed via New Malden and Kingston entering the sidings at the Hampton Wick end, and trains of such vehicles returned to Waterloo via Strawberry Hill and Twickenham leaving the sidings at the Teddington end with no use of Clapham Yard being made for these trains. Immediately before the infamous month-long RMT guards' strike on South Western Railway in December 2019, I had a ride round the Kingston loop to establish whether any remains of the sidings were still visible; they weren't, with modern housing seemingly dominating the site.

Provision of empty vans. When perusing Christmas Mails and Parcels Traffic notices, one is struck with the seemingly large number of entries where specific stations had to "arrange" or "secure" vans (to use the notices' terminology) as necessary to provide for special workings. It is easy to imagine frequent – and maybe sometimes hectic and even colourful conversations between Station Masters and their respective passenger rolling stock offices which were located at Waterloo, Eastleigh and Exeter Central, to arrange van provision: nevertheless, P notices did include some empty van workings. To meet the heavy demand for vans over the Christmas period, quite extensive use was made of vanfits, some of which were pre-labelled "Reserved for Parcels Traffic".

Analysis of individual years

1952 Notice No. P73, LWD, 1952, operative Sunday 14 December to Saturday 27 December inclusive. This notice together with most or all others were divided into five sections:

1. Alterations to booked services.
2. Additional Parcels and Mail Trains.
3. Altered formations and revised loading instructions.
4. Additional vehicles by ordinary trains.
5. Freight train alterations. By 1956 and onwards, however, freight train alterations were no longer included, such details being included in subsequent notices.

Following on were a list of signal boxes where revised opening hours, principally extended hours, applied affecting the following boxes: Teddington, Bookham, Horsley, Clandon, London Road (Guildford), Cobham (extended for passing of one train only!), Winchester Junction, Lyndhurst Road (as Ashurst, New Forest was then called), New Milton, Sutton Bingham, North Tawton, Coleford Junction and Newton St Cyres. In marked contrast Oatlands, located between Weybridge and Walton-on-Thames and not normally open continuously for 24 hours, had to remain open continuously from Monday 15 December until Wednesday 24 December. This was necessary because much use was made of the adjacent carriage sidings for berthing and forming additional van trains. Meanwhile other boxes had to be opened as arranged by local Station Masters for passing of trains shown in the notice.

91

Provision of shunting engines

The sight of a relatively small shunting engine was quite common at larger stations and yards, pottering about here and there positioning vans and maybe passenger carrying coaches into correct formations for the next train they were to form. But quite clearly those stations dealing with vastly increased amounts of mails and parcels traffic over Christmas periods would become inundated without additional shunting power. To meet these requirements it was necessary to ensure adequate motive power was available rostered for "coaching shunting" which was the official term for carriage shunting. For this additional engines were sometimes diagrammed or hours of shunting extended or revised. Whilst space does not permit full details of revisions at individual stations as shunting hours varied quite widely according to station requirements, the following is a summary of stations affected: Waterloo, Nine Elms (Hay Yard), Woking, Hounslow, Portsmouth & Southsea, Fratton Old Yard, Teddington, Walton-on-Thames (Oatlands sidings), Aldershot, Guildford, Basingstoke, Southampton Terminus, Salisbury (2 engines), Ash, Barnstaple Junction and Reading South.

Walton-on-Thames and Teddington

It should be noted that no loading or unloading of vans forming special trains was undertaken at these two locations, which were staging points for re-marshalling of trains proceeding onwards to such destinations as Nine Elms 'J' Shed. So now continuing the year 1952.

Special Instructions

Details of signal box opening hours extensions and provision of shunting engines have already been given and this is followed in the notice by a plethora of special instructions abbreviated as follows:

1. All letter mails, livestock, valuable consignments, perishables, evergreens, market and "To be Called For" (TCF) traffic for London delivery or cross-London transfer had to be loaded to Waterloo by ordinary trains and not by parcels specials. Prior arrangements had to be made for complete van loads.

With Walton-on-Thames station in the background, BR Standard Class 4 No. 75075 passes on the down fast line. A shunting engine was provided near here at Oatlands sidings. *S/Way Collection*

Christmas Mails and Parcels Traffic

2. All parcels post and ordinary parcels traffic for London delivery or cross-London transfer were, except where special provision was made, conveyed by special trains or vans and dealt with at Nine Elms and were not to be forwarded by ordinary trains to Waterloo. Stations not served by special trains or vans for this type of traffic had to use ordinary services to the nearest point served by special trains for transfer there.

3. London suburban stations not served by special trains to use ordinary services scheduled to arrive Waterloo before 4.30pm and after 7.30pm.

4. Separate vans were provided on special trains to Nine Elms for (a) parcels post and (b) ordinary parcels traffic. It was of the utmost importance that vans with category (b) traffic arrived at Nine Elms next to engine and category (a) traffic at the rear.

5. Parcels post despatches from London were segregated as follows, (a) to be dealt with at Waterloo:- suburban (including Vauxhall to West Byfleet; West of England (West of Basingstoke); Channel Islands. (b) to be dealt with at Battersea;-;Woking to Weymouth (inclusive); Woking to Portsmouth and Isle of Wight; Woking to Alton.

6. Schools Luggage in Advance (PLA). Dealt with at Nine Elms, stations to label vans accordingly and forward by special parcels trains where possible. This was Boarding Schools' Christmas holiday traffic.

7. Parcels Post and Letter Mail Traffic. If additional vans had been required over and above those specified in the P notice, Station Masters at stations concerned had to report the matter to their respective District Superintendent before 29 December, detailing vans provided and stating if the extra vehicles were actually ordered by postal authorities and, if so, enclosing the written request received.

8. A reminder that throughout the period covered by the P notice, vans should not be allowed to remain idle at stations. Should any be received and not utilised within 12 hours of arrival, details to be telephoned to the Superintendent of Operation Rolling Stock section or local stock office.

9. Guards' brake vans to be used economically with only one provided in each special train, except where a train is booked to divide en-route when it is essential a brake van be formed in each portion. Correct provision of brake vans

Oatlands sidings, on the up side west of Walton-on-Thames, is one of those locations where few images seem to have been taken. Here though Keith Lawrence has captured what was fast becoming a somewhat overgrown location on 4 July 1966. Passing on the down fast is No. 34019 *Bideford*. *Keith Lawrence*

was of paramount importance and Supervisors were to give this matter particular attention. In any instance where Corr PMVs were booked but unavailable, one PMV(4) should be provided in lieu. Where bogie guards' vans were booked and unavailable, one PMV(4) and a 4-wheeled guards' van were to be provided.

10. Van labelling. Vans conveying ordinary parcels traffic to be labelled "PARCELS TRAFFIC ONLY". Vans conveying GPO Parcels Post to be specially labelled "PARCELS POST ONLY" with labels placed in a low position at leading end of vehicle. All parcels vans formed on special and ordinary trains from Waterloo and Battersea to bear a label on van windows each side detailing stations for which parcels are loaded. Windows to which labels are affixed to be kept closed. When vans are cleared of traffic, destination stations to arrange label removal before vehicles released for further use, to which Station Masters had to give special attention.

11. Special electric parcels trains. These trains had to carry headcode U or 97. 1952 was, of course, still in the age when suburban trains were still formed of pre-war stock which carried letter headcodes and Bulleid stock which displayed numerical headcodes.

12. Formation of trains to Nine Elms "J" Shed. Loads of these trains were not to exceed = 35 wagons in length (700 feet), which Walton had to note and arrange accordingly.

13. Use of vanfits. To assist in catering for the heavy demand during the operational period of the P notice, a number of vanfits (labelled "Reserved for Parcels Traffic") were supplied for passenger trains van use. The P notice did not specify who had to affix these labels, maybe the Freight Rolling Stock section undertook this responsibility.

14. The through parcel van working to King's Cross was withdrawn for the notice's duration. Presumably such vans worked via the LTE Widened Lines but this is supposition. Instead, traffic was dealt with at Waterloo but for about a week before Christmas was transferred from Waterloo by road to King's Cross Goods Station. Vans which would have worked through to King's Cross were provided as usual and labelled "Nine Elms for King's Cross". All concerned were instructed to give the instruction special attention.

15. Horse Boxes were not to be used for loading mails and parcels traffic.

Van train from a different era; empty vans leaving Basingstoke west yard for Clapham Junction behind Class 73 No. 73142.

Christmas Mails and Parcels Traffic

16. Newspaper and Stove Vans were not to work off the Southern Region, and had to be reported to Waterloo Rolling Stock Section if found on hand or not on booked working.

17. Vulnerable Traffic and Security Bags covered by waybill not to be loaded into vans dealt with at Nine Elms but continue to pass via Waterloo on normal services.

18. Parcels Post to Birmingham and Birmingham transfer. During the Christmas Mails and Parcels period, through van loads of Parcels Post for Birmingham and Birmingham transfer were dealt with at Birmingham (Monument Lane), but that station could only deal with four-wheeled freight vehicles. Stations had to arrange loading of this traffic into that type of stock.

Additional Van trains run

So now to the summary of additional van trains run in 1952 which fell into three categories: Weekdays – Mondays-Fridays shown as (WD) after title, Saturdays (Sa) and Sundays (Su). Clearly space does not permit detailing actual dates individual trains ran, so they are shown as days running by listing them into one of the aforementioned categories. Once again a reminder that it should be remembered that these trains ran each year, so in following years only principal alterations will be shown.

Waterloo-Battersea-Woking-Bournemouth and Weymouth

12.15am (Su) Eastleigh-Southampton Terminus

6.15am (WD) Battersea-Clapham Junction

6.15am (Su) Battersea-Woking via East Putney

9.45am (Su)/9.52am (WD) Clapham Junction-Wimbledon West Yard via East Putney thence to Wimbledon Volunteer Siding as arranged by Station Master, Wimbledon.

2.15pm (WD) Battersea-Woking via East Putney

2.15pm (Su) Battersea-Clapham Junction with vans of parcel post only for Woking to be attached to 3.35pm empty milk tanks Clapham Junction-Exeter Central for detachment at Woking.

3.2pm (WD)/5.45pm (Su) Waterloo-Weymouth 4.0pm (WD) Eastleigh-Southampton Terminus 5.15pm (So)/7.45pm (WD) Battersea-Woking via East Putney

8.0pm (WD) Waterloo-Bournemouth Central

6.20pm (Su)/6.45pm (WD) Weymouth-Walton

10.14pm (WD) Dorchester South-Eastleigh

11.10pm (WD) Woking-Walton

Part 2 will appear in SW56. For reasons of space we regret 'Rebuilt' has also had to be held over.

Parcels in the 'four foot'. A commonplace sight for decades with the porter trackside and either waiting to cross or having already crossed perhaps. The parcel will be noted. The train is the 3.51p.m. Eastleigh to Fawley formed of WR stock (off a Didcot – Eastleigh) service and with the addition of what appears to be a PMV on the rear.

The Southern Way

The regular volume for the Southern devotee
MOST RECENT BACK ISSUES

The Southern Way is available from all good book sellers, or in case of difficulty, direct from the publisher. (Post free UK) Each regular issue contains at least 96 pages including colour content.

£11.95 each
£12.95 from Issue 7
£14.50 from Issue 21
£14.95 from Issue 35

Subscription for four-issues available
(Post free in the UK)
www.crecy.co.uk

96